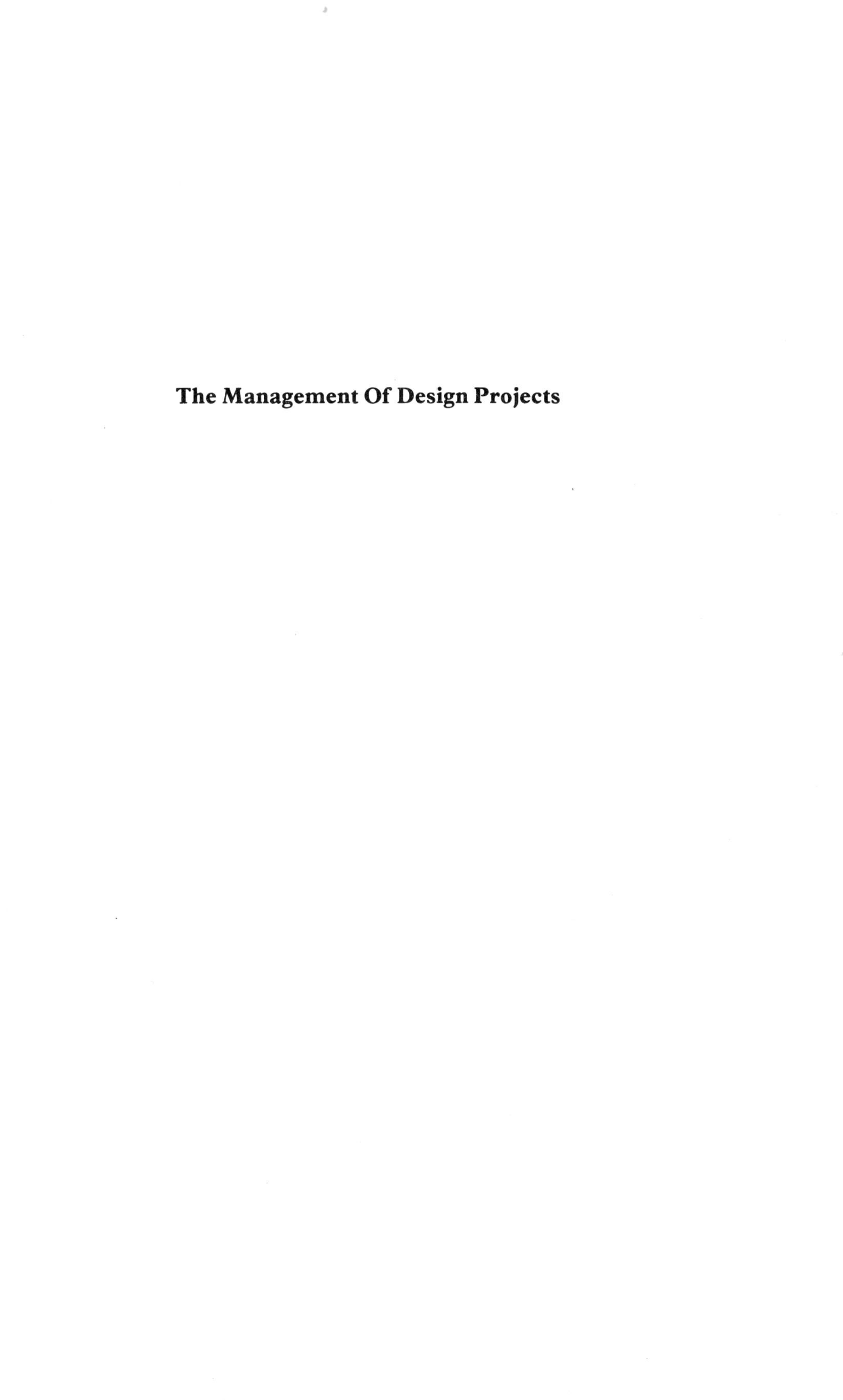

The Management Of Design Projects

The Management Of Design Projects

Alan Topalian

Associated Business Press
London

Published by
Associated Business Press
An imprint of
Associated Business Programmes Limited
Ludgate House
107-111 Fleet Street
London EC4A 2AB

First published 1980

ISBN 0 85227 078 X

Typeset in 11/11 Plantin by
Photo-Graphics, Stockland, Honiton, Devon
Printed and Bound in Great Britain by
Biddles Ltd, Guildford and King's Lynn

Contents

This book is dedicated to those managers who believe that design has little to contribute to the profitablity of their companies. For it is they who are forcing us, in the design professions, to think more carefully about our roles in industry and to explain these roles more clearly.

Foreword

The nineteen-seventies will be remembered by many as a time of reality: a time when the world could no longer ignore the far-reaching changes in technology and social behaviour that had mounted over preceding decades. It will probably be remembered too as a time of much talk but little effective action — when old problems increasingly failed to respond to old solutions. It was in this climate that design became a popular subject for discussion, thrown up by Britain's continuing poor export record.

We are all too aware of the way we lag behind in design and development compared with our principal foreign competitors, yet Britain can claim the largest and most highly trained body of designers in the world. Many of the well-designed foreign products we buy were originated by the very British designers who are grossly under-used at home. At the core of this apparent paradox is the fallacy that the design process is alien to industrial management.

While all the talk and head-scratching has gone on — by industry, government and the design profession — one man has persistently championed a simple truth. That man is Alan Topalian, and his simple truth is that *it is the management of design, not the process of designing, that is alien territory to British managers.*

The traditional criticisms of design as an industrial discipline run as follows: 'Design is only superficial; design is subjective and unsuited to industrial management; there is no way of evaluating the results against the investment; designers are an expensive overhead; it is difficult to communicate with designers; there is a lot of bad design about and so there must be lots of bad designers about as well.'

Of course there is an element of truth in all these arguments, so industry has felt comfortable in its ignorance and inactivity. The real truth emerges most clearly in this book.

Alan Topalian maintains that design must take its place in the mainstream of industrial investment, and that the management of design investments involves considerable skill; he points out that there are ways of evaluating the results of design investments that need be little different from other forms of business evaluation and that design decisions, like all other decisions, can be subjective but they can also respond to rigorous analysis. Those who believe that design can make only fleeting contributions to business profitability clearly need to be disabused; a professional designer, far from being an artist solely concerned with creating personal expressions, is actually trained specifically to create appropriate and commercially viable solutions to business problems.

Designers know that in common with all specialisations, language difficulties exist, but they are exaggerated, and jargon on both sides can easily be avoided or explained. This book is written in a language which will be familiar to managers, and rightly so, because it is principally addressed to management — not as a complaint but as a realistic guide to how designers might be used more effectively in industry. Without apology, it repeats its main message: that the project analysis and problem-solving aspects of design are absolutely in character with those already used by industrial management. Clearly, managers who deal with design will find the book invaluable.

The design profession is already indebted to Alan Topalian for his championing of this cause, and now for his book. And it's interesting to note that he is producing results without financial funding in a field which has received little sponsorship since it hit the headlines a few years back. Sadly, where funds *have* been invested, it would seem that there has been a notable lack of progress.

Coffee tables are piled high with design publications and, of course, books about design are usually picture books. This book is essentially different in that it aims to help solve problems rather than merely to stimulate or provoke. So it requires reading and, in my view, *must* be read by all levels of management that come into contact with the design process — that is, if our industry is to keep pace with change.

Richard Negus PPSIAD, FSTD

Introduction: Starting almost from scratch

All 'outputs' produced by organisations are designed. This is true whatever the outputs, be they products, services, packaging, exhibition stands, promotional literature, interiors of shops and offices, or stationery. The quality of any output is dependent on the resources allocated to its production, particularly the design and other specialist skills employed, and the manner in which the design project is managed.

Experience suggests strongly that design standards are influenced far more by the way design projects are managed than by the design/specialist skills involved. Experience also suggests that, whatever design policies are agreed at board level, it is during the day-to-day management of design projects — typically entrusted to those much lower down the management hierarchy — that the design standards of an organisation are established and maintained.

Whereas the veracity of these statements may not be self-evident, management's ultimate responsibility for the outcomes (commercial or otherwise) of all design projects must surely be beyond dispute. Management, within both the private and public sectors, represents the major market for designers' skills. In controlling what is produced, management controls which designs get to the market place — even allowing for the influence of designers, market forces, and statutory constraints.

Thus quality of design ought to be a prime concern of management, but all too frequently it is relegated to being a burden of the design professions. Given that this nation's prosperity depends so much on the effective design of manufactured goods, it is surprising that no serious, sustained attention has been paid to the way design resources and design projects are managed.

This book sets out the texts of eight lectures on the nature, management and evaluation of design projects — specifically those projects concerned with product, graphic, exhibition and interior design. The lectures derive from a formal research study which was launched in September 1976. In setting up this study, three commitments were specified: first, that management should be the primary target audience; second, that results should be published quickly; and third, that teaching material should be generated for both management and design 'students'.

At the start of the study, extensive discussions with designers and managers indicated that, on the surface, there is little common ground between them — either in viewpoint or language used. This was anticipated. What was not anticipated, however, was the finding that there is little common ground amongst designers when discussing their work.

Increasing understanding between designers and their clients — arguably a prerequisite to the more effective use of the design talent available — becomes more difficult if no common language exists for discussing design projects. Consequently, the main thrust of the research study over the past two years has been the development of a conceptual framework by which the nature and management of design projects could be analysed in depth, and taught.

The framework is essentially a first-stage distillation of the experiences of designers and clients across a representative range of design projects and the structuring of these into an accessible body of shared experience. As a result of the articulation of the framework and the promotion of the common ground, a 'language' is emerging which will enable more detailed research to be carried out.

No apology is made for the basic approach that has been adopted in the discussions. Many of the points may be obvious — obvious, that is, *after* they have been articulated. Many of the arguments put forward need to be developed from 'first principles' which are frequently taken for granted during discussions on design but are rarely articulated in a way that can be tested — either as to their substance or for agreement between the parties in discussion.

As there appear to be no data on how managers view their contribution to and role in design projects, a survey was devised, aimed at determining what managers/clients perceive to be the problems when managing design projects. It was hoped that apart from generating valuable data, a demonstration of concern for management difficulties might encourage managers to join in serious debate on 'design management'. The findings of this survey — the first of its kind in this field — suggest that far from being critical of designers, managers admit to a disturbing unease about their understanding of, and skills in handling, design projects. The findings of this survey, and those of a complementary survey of designers' perceptions, are given in Appendix A.

These are early days for 'design management'. Hopefully the following lectures — which make up a basic course in the field — will help improve the understanding between designers and their clients.

1 Meaning, control and the design process

Design is a principal source from which structure and meaning are derived. Any arrangement of parts, forms, materials, colours, and so on, constitutes a design. Nature is a storehouse of such arrangements evolved through a series of chance events over millions of years as means became adapted to ends. A great many of these designs are constructed by primitive creatures during the struggles for survival which form their lives. Virtually all their activities are instinctive: that is, there is no conscious intention. Furthermore, they are strictly limited from creature to creature as each has adapted to fit closely into its particular environment.

By contrast, Man evolved a greater flexibility of existence in that he is not locked into any specific environment. When we talk about Man's design activities over the ages, conscious intervention in natural events and purpose are key issues. Though Man will instinctively explore his environment, he does not inherit any ready-made design skills but has to develop them through life. Man is unique among living creatures in that he has evolved the capacity to design through conscious effort. With this imagination he has combined his abilities to invent and, through invention, make discoveries. With succeeding discoveries and inventions, Man's abilities and grasp of his surroundings were extended: increased knowledge enabled him to handle larger quantities of information; multiplying skills brought together in more complex combinations have produced achievements of progressive intricacy and widening impact. Thus Man's uniqueness is further demonstrated by his capacity to produce a wide range of designs. Clearly, in certain circumstances, his need to survive demanded greater control over his environment and his design efforts were directed towards that goal. In other instances, Man's design efforts amounted to little more than stamping his presence on the environment: a celebration of his presence through a representation of himself and his skills, and an extension of it after death. Gradually Man shaped his surroundings, exerting more and more control, till he gained dominance over the earth.

Through the ages, nature has provided Man with an 'order' of things around him, and Man's perceptions of himself and his environment have been influenced by this 'order'. As his comprehension of it altered — sometimes spurring his inventiveness on, sometimes constraining it — a natural conception of 'rightness' and 'wrongness' developed. Indeed, history can be interpreted as a record of Man's perceptions of himself, his fellow creatures, his particular environment, and his place within that environment: his deeds and the artifacts he produced were all heavily influenced by these perceptions. Today, though Man's design skills may be highly developed, perceptions exert no less an influence on designs produced. And in such perceptions can be found the sources of some of the emotions with which we react to different designs.

Thus it could be argued that the *process* of designing is a process by which structure and meaning are introduced into the world. Yet when we look at the mounting visual pollution around us, it is clear that most design activity makes little sense. Why is this so? Well, one explanation must be that physical dominance without understanding is a recipe for disaster. Structuring the environment and endeavouring to invest it with meaning are not enough: *effective* control in the long term will only be achieved through better understanding, for it is understanding which makes the environment truly accessible to Man and allows him to live in harmony with his surroundings.

Design and the satisfaction of needs

The process of designing ought to be directed towards the satisfaction of needs or, more precisely, the development of means by which needs might be satisfied. The needs in question are principally those of Man — be they physical, cultural, legal, political, etc. — and environmental needs which are perceived to affect him.

Clearly, management interest might be expected to concentrate on those needs that can be satisfied by the purchase, use and consumption of goods and services: those needs that are reflected in the market place, predominantly through demand. Furthermore, it will probably be difficult for managers to ignore the needs *within* their organisations, the satisfaction of which will be reflected in harmony amongst employees and in efficient production. Clearly, too, management might be expected to concentrate on those aspects of the environment to which their organisations are particularly sensitive and on those which they can influence profitably.

In reality, a great deal of design activity is not directed at the satisfaction of needs in the longer term, but at pandering to short-term 'wants'. And to make matters worse, within this class of design activity, the 'wants' derive frequently from managers and designers, *not* from the final consumers for whom it is intended. Industry still finds it easier to harness its forces to encourage new 'wants' rather than to finding out about existing needs. Designers, too, often find it easier (and more enjoyable, of course) creating solutions *they* want to produce rather than solutions which accord more closely to client and end-consumer needs. This 'production orientation' — by which firms concentrate on selling what they can make rather than organising themselves to make what markets need — is clearly a legitimate strategy to adopt. However, it represents a more difficult marketing proposition, as space has to be created for such outputs within the already crowded spectrum of market demands. If firms which are production orientated are also deficient in their marketing (which is frequently the case), then one difficulty is compounded by the other: they produce outputs which cannot be made attractive enough to the market.

It should be stressed that a strategy of production orientation need not be doomed to failure; far from it. For many firms it has obviously been very successful, either because of happy circumstances (where, for example, what was produced happened to coincide with what was needed), or because a company genuinely leads market trends.

Nevertheless, managers and designers do go seriously astray when they begin to believe they can be effective arbiters of consumer needs without researching such needs sensibly; when, in ignorance or sheer arrogance, managers and designers begin to rationalise their private wants into fictitious or superficial consumer needs. Without knowledge and understanding of consumer needs, production orientation may be rationalised into 'marketing' orientation; 'leading' can be transformed — through a series of imperceptible steps — into 'imposing'. There is no question that this activity — which, at best, has as its basis the superficial interpretation of needs and, at worst, their wholesale neglect — is 'design' activity. What should be made clear, however, is that from a professional management point of view, this class of design activity makes little sense because it does not help to ensure profits in the long term. If managers seek to maximise the chances of generating profits from the design work they commission, they should ensure that all such work has as its foundation the accurate articulation of real needs into properly structured and meaningful problems. This, in turn, should lead to a deeper understanding of those needs as well as to an enhanced awareness of the effective means that might be used to satisfy them. And in awareness and understanding lie the seeds of effective choice and control. From the professional management point of view, therefore, the chief contribution of design to business profitability should be sought in the structure, understanding and control it brings about during the process of satisfying needs. But with increased control comes greater responsibility which should mean that managers progressively put more care into the design activity they oversee.

However, this conception of design is dramatically different from that which is commonly held by managers, for it would appear that many managers associate the principal contribution of design (and designers) with the cosmetic packaging of outputs produced — as though one had to 'sugar the pill' all the time. Relatively few managers see the necessity to involve designers at the outset of design projects. It is not difficult to see how this attitude to design and designers has come about. Where industry develops with a tradition of production orientation, managers tend to know what they want to produce: they do not need assistance in deciding the nature of the outputs they produce, hence they do not feel the need for assistance in asking the right questions. By contrast, they are fully aware that what they produce has to be made presentable to the market — at least as attractive as anything which is not needed can be made so. In essence, cosmetic packaging — which has turned out to be a powerhouse of pollution.

A definition of design

Design projects involve the control of the design process as well as the 'production' of designs. When we talk about *the design process* what exactly do we mean? As yet there is no universally accepted or comprehensive definition of this process. Because creation can be such an individual act, there are a multitude of ways of designing. Some processes may carry greater chances of resulting in successful outcomes than others, and some processes may be more efficient than others. But no one process of designing carries with it a guarantee of success, irrespective of the problems and people involved.

But when analysing the many and diverse guises of design activity, the one constant that stands out is the development of means to ends. Whether this activity is spurred on by a dissatisfaction with what exists or by a vision of what could be; whether the stimulus derives from existing opportunities in the market (market-pull) or from the opportunities that might be created as a result of inventions (invention-push), makes little difference: ultimately, design is about the development of means to ends. More precisely, *design is the process by which needs in the environment are conceptualised and interpreted into instruments which are formulated to satisfy these needs.*

According to this definition, design is inherent in most activities carried out by industrial and commercial organisations. Products and services are designed, as are corporate strategies; publicity material and information systems are designed; productivity deals and company images are designed; research programmes are designed, as are working environments such as shops, offices and factories. Any series of decisions which lead to means being adapted to ends are examples of the design process, even though many of these examples have no *visual* input or manifestations. Therefore, a competence in design represents an essential tool for management.

This definition covers the fundamental stages in the design process: conceptualisation, interpretation, and formulation. Whether designers or their clients are conscious of it or not, they go through these stages

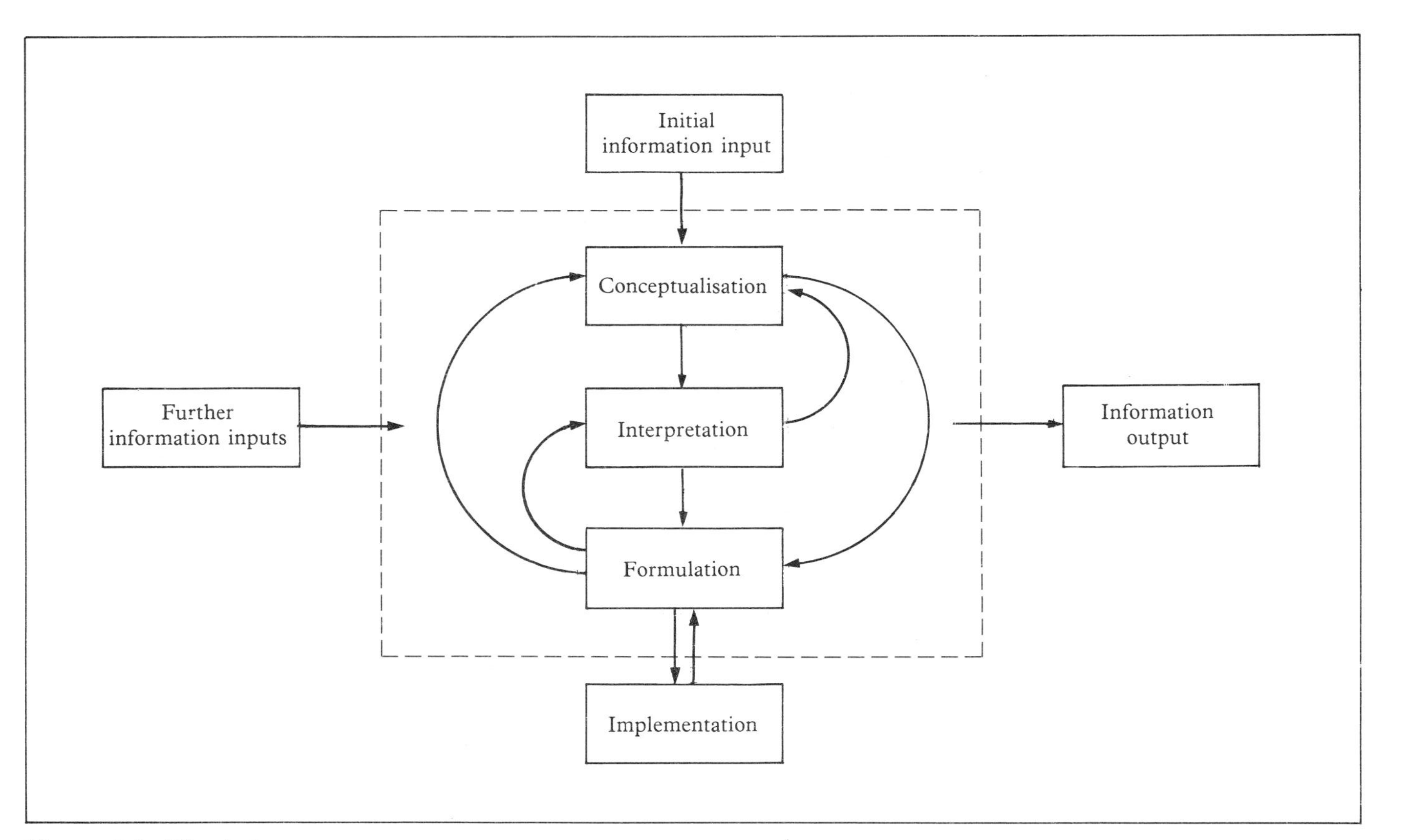

Figure 1.1 *The design process*

individually or jointly — though perhaps not strictly in that specific one-way sequence.

Interestingly, these stages apply equally to the 'design', or definition, of problems as to the 'production', or design, of solutions. Problems — just like solutions — need to be conceptualised, interpreted and formulated. There is nothing God-given or 'obvious' about them. And as the way problems are perceived can have a dramatic impact on the chances of developing effective solutions, the design process must have as much to do with finding the right questions to ask (in the choice and structuring of problems) as it has with providing appropriate answers (in the formulation of solutions). Indeed, defining problems can demand greater originality than the formulation of solutions. Witness the fact that we rarely know how to exploit our knowledge and technologies to the full: today we have more solutions floating about than we perceive problems to which they might be applied.

This definition also provides an insight into the complete range of design activity which might be encountered. At one extreme, where design activity demonstrates the greatest originality, the design process starts with a search for novel conceptions of needs and proceeds towards the formulation of novel solutions. At the other extreme, where design activity is at its most pedestrian, common solutions are produced in response to well-known needs. Between these extremes, there are grades in originality. New need concepts can be discovered, only to be interpreted and formulated into relatively unimpressive solutions; or startlingly new solutions might be conceived and formulated for well-known needs. Originality may be demonstrated throughout the design process or only within certain stages: in conceptualising, interpreting, or formulating either the problem or the solution.

These considerations should have a critical impact on the evaluation of design activity and solutions, particularly when the respective contributions of designers and clients to outcomes are being determined. For on some occasions, clients will value the fresh viewpoint designers can adopt when they are allowed to seek out problems, and designers may be involved almost single-handed in the definition of problems right through to the formulation of the solutions. Those would be fully-fledged design projects.

On other occasions, the designer will only be asked to formulate a solution to a tightly defined problem and solution concept; the client may actually dictate the form of solution, in which case the designer's involvement relates to the translation of the solution in the client's mind into physical reality — no more, no less. These would be 'pair-of-hands' assignments. It is normally at this end of the spectrum that designers get involved in visual design at the most superficial level — that is, cosmetic packaging. The contribution made by the designer and hence the influence he has on the outcome of a design project

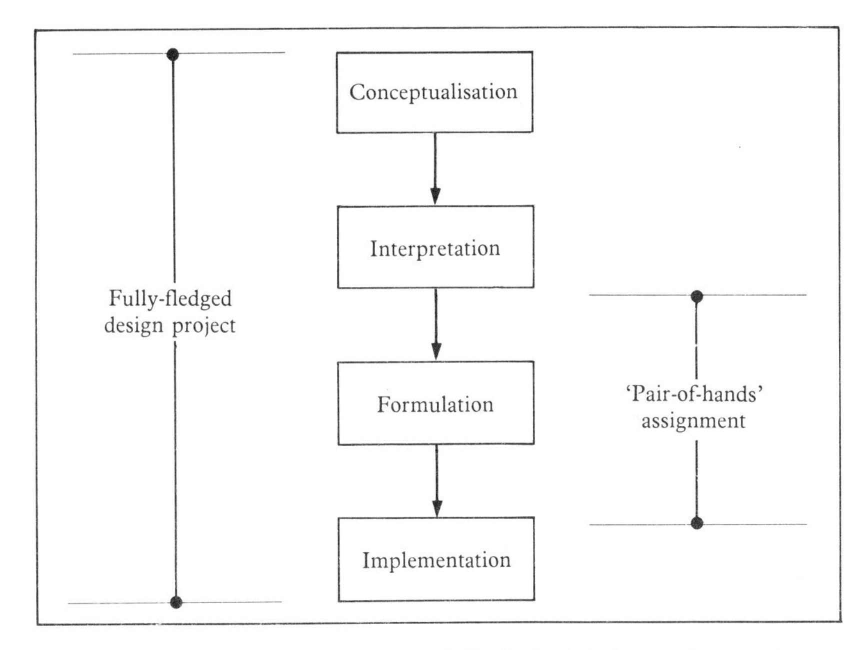

Figure 1.2 *Designer involvement in fully-fledged design projects and 'pair-of-hands' assignments*

obviously varies from a maximum at one end of the spectrum to a minimum at the other. Yet because the design process tends not to be analysed in these terms, these considerations are rarely taken into account during project evaluation.

Different ways of designing

It is not surprising that there is no common perception of the design process, because there is no one way of designing. One designer will proceed with his work in a different way from another. Frequently, designers adopt different design procedures when handling different types of design problems. To make matters more complicated still, a designer might adopt a different procedure when tackling the same problem a second time — not necessarily because he learned anything the first time round or circumstances dictate a change, but simply because he feels like it. This, of course, is no different from the way managers might tackle business problems. But what are the different procedures designers go through in order to solve problems and satisfy needs?

The most obvious procedure is to start with a clearly defined problem and to work logically through to the formulation of an appropriate design solution: this is a 'natural' problem-to-solution sequence which involves understanding the problem properly before even toying with tentative solutions. It involves formulating the right questions to ask and agreement on what decisions have to be made during the design process. Yet it is frequently criticised as being not quite a relevant model within the context of design problem-solving because it is argued that design problems cannot properly be defined or grasped intellectually without some reference to possible or sought solutions. Furthermore, reference to one set of solutions as opposed to another effectively changes the definition of a design problem.

Clearly it is not always possible to define problems accurately (or even to ask the right questions) at the start of projects: there is just so much that can be known about certain kinds of problems before work starts on solving them. Paradoxically, too, there is the danger that too much stress on the definition and interpretation of problems can blunt the enthusiasm of those involved to grasp at solution opportunities. But there is also a potential conflict in interpretation: does a performance specification for a product, for example, constitute part of the definition of the problem or part of the solution? For a brand-new product concept, the drawing up of a performance specification might well be considered part of the solution; where an existing product is being upgraded, a similar specification would almost certainly be considered part of the definition of the problem. There is a class of problems for which definitions are also specific, sometimes unique, prescriptions of the form of solutions. Outside this class, however, a specification of solution would not, *on its own,* constitute an adequate definition of a problem.

Another common design procedure might — when practised competently — be termed the *'intuitive leap'* process: the designer lights upon a fairly clearly formed solution without much analysis of the problem. 'Designing' in these cases does not relate so much to the *conception of a new solution,* as to *'fitting' a pre-formed solution* to the problem. It is sequential but proceeds backwards, from solution to problem. Designers argue that years of experience and 'flair' contribute to each 'intuitive leap'. Certainly this is true in many cases; design proposals put forward fit closely with the problems, even to the extent of providing completely new insights into their nature and how they might be handled. In other cases, however, a 'leap' will represent little more than a veneer for intellectual apathy: tired, 'standard' solutions are stretched to deal with problems for which they were never originally intended. Furthermore, most designers have repertoires of solutions and constantly look out for appropriate problems to which they might apply these 'floating' solutions.

A third common design procedure is to work, as it were, from both ends: alternately working with the problem and with the solution; partly conceiving brand-new solutions, partly 'fitting' elements of familiar solutions into new arrangements until an appropriate solution is reached. This is essentially a combination of the first two procedures whereby the designer switches between procedures as he grapples with the problem. Even though he switches, the designer may still, on balance, be problem-biased or solution-biased in his approach.

Whichever procedure the designer adopts during the course of a particular project, he may approach the work in different ways. At one extreme, there is a 'mechanistic' approach. The programme of work is set out in a strict, one-way sequence. The problem is seen as clearly articulated and fixed. At each stage, the decisions made are aimed at reducing the options and uncertainties in succeeding stages. In the ideal project, the progressive 'firming up' that occurs would lead to a single, most appropriate solution — a very rare occurrence indeed! At the other extreme, a fluid 'organic' approach may be adopted. Here the stages of work are not necessarily taken in a programmed sequence: several stages may be worked on concurrently, and solutions are evolved through an iterative process *between stages.* It is recognised that problems can be defined in several equally clear and plausible ways, and that problems do not necessarily remain unchanged once they have been defined. Thus in adopting a fluid approach, the designer alternately explores, refines and even re-positions both problem and solution. Various layers of information will be gathered as progressive layers of problem and solution are examined. Tentative decisions will be taken and the implications tested across a span of project stages through to implementation. Several options of solution may be pursued in parallel to varying degrees of development, with continuous evaluation, before the final decision is made on a workable problem and a corresponding solution. Thus decisions are not firmed up until they absolutely have to be, whereupon a particular *series* of decisions might be confirmed.

Depending on how experienced the designer is, he will deal with details or elements of problems, or combinations of elements. The more experienced the designer and the more familiar he becomes with the problem, the more able he should be to manipulate progressively larger combinations of elements. However, it would be wrong to assume that the progression is naturally 'one way'. The 'mechanistic' design problem-solving approach might suggest that this is the case: a succession of decisions concerning various details, then aspects, of a problem build up to a solution for the problem as a whole. This certainly works with some design problems, but not all. Effective solutions to constituent elements of a problem do not always lead to an effective, integrated solution to the problem as a whole. Consequently, the ability to structure problems sensibly is of paramount importance, especially so that the interrelationships between elements, and between elements and the whole problem, can be understood. But as details of problems do not always reveal themselves before the problem is grasped as a whole, it is often equally important to be able to deal simultaneously with the details of a problem and with the problem as a whole. The broader the span of levels a designer can handle, the more flexible he is likely to be in the design procedures he uses.

Whichever combination of procedure and approach the designer adopts during the course of a particular project, he may go about the work in a variety of ways. Thus, in working from a definition of a problem forward to the conception of a solution, there may be a 'natural' sequence to his programme of work: analysis of problem, determination of opportunities and constraints, generation of alternative solution concepts, evaluation of solutions concepts against project objectives and criteria of acceptability, choice of solution concept, detailed formulation of solution, and implementation of solution. Nevertheless, the circumstances surrounding the project may not allow this sequence to be followed. It may be known, for example, that vital research findings will not be available at a given point in the project; certain design decisions may be required earlier than when they should ideally be made because the client will be out of the country for a time and the project cannot be delayed. Thus the *feasible* programme of a project may vary from the ideal programme — in content, sequence, and timescale.

In other circumstances, the designer may programme his work along a line of 'least resistance' or 'greatest familiarity'. In other words, he picks his way through the problem-solving process by dealing first with those aspects which are, to him, easier or more familiar — perhaps in the hope that as the project progresses, the difficult and unfamiliar aspects will somehow fall into place or seem less daunting, or that adequate compromises suggest themselves. Client resistance to unfamiliar solutions may also influence the way the designer goes about his work: sometimes the difficulties encountered in 'selling' a novel solution outweigh the benefits which would derive from being different. Alternatively, the designer may prefer to start work on a problem by tackling its critical or more difficult features — sticking with these until acceptable decisions are reached or treatments devised. The philosophy behind this approach is not that the principal battles in design projects derive from the critical features (for this is not always true), but rather that one tends to learn more about problems when tackling their critical features — in this way one can get to grips with problems faster. All things being equal, the more accurate the diagnosis of a problem early on in a project, the greater the potential for devising an effective solution; the greater the understanding of a problem early on in a project, the greater potential for planning work ahead and hence control: altogether, the greater the chances of achieving a successful outcome.

Orientation in design problem-solving

'Orientation' in design problem-solving has already been touched upon — whether the designer proceeds 'forwards' from a given problem to conceive a solution, or 'backwards' to fit a largely preformed solution to the problem. However, designers frequently discuss a different kind of orientation: for example, whether they should place greater emphasis on 'content' (what it is they are designing: the material design) or on 'presentation' (fundamentally its visual design). Should content dictate presentation? If so, should any work be attempted on presentation before content has been completely specified?

Alternatively, is there any merit in presentation dictating content? And in what circumstances might such an orientation prove successful? Would it, for example, be acceptable to deal with the visual design of a promotional brochure before a word of copy has been written? It can be done, of course; quite a few managers demand it, and many designers are happy to undertake work in that order. But is this order 'correct' and is it likely to result in effective solutions? Again, should the exterior of a shopping centre always *reflect* the interior atmosphere, or would it be acceptable for the exterior to *set the tone* for the internal treatment?

Put simply, should designers design from the 'inside' outwards (that is, from content to presentation), or should they design from the 'outside' inwards (that is, fix the presentation, then sort out the implications for the contents)?

As with arguments for or against individual design procedures, arguments about orientation have largely lost their relevance today. Just as there is rarely one right solution to any problem, there is rarely one right way of designing to the exclusion of all others. Though it is frequently necessary for content to dictate presentation, every now and then there are positive advantages in presentation dictating content. But an assumption underlies these questions which is rarely challenged: that content and presentation should always be related visually for design solutions to be successful; and experience certainly bears out that, unless content and presentation *are* coordinated in an 'acceptable' way, a design solution is unlikely to be taken seriously. Unfortunately, worrying too much about 'acceptable' associations between content and presentation can lead to designs which are devoid of imagination, for instead of exploring opportunities which might lead to fresh solutions, the search is confined to familiar and safe territory.

There is no reason why unusual associations should not be explored to get a better understanding of what represents 'acceptability' as well as novel perspectives on the problem. For example, an office designed to look like a supermarket might, superficially, strike one as a silly proposition. But is it? What is wrong with exploring the concept of an office as a supermarket where a great variety of information is 'piled high' and is quickly accessible; where the disposition of information sources is such as to foster quick processing and decision-making, all set within a bright and scrupulously clean environment? Similarly, a home computer packaged as Yogi Bear sounds distinctly odd. But what is wrong with exploring the concept of a home computer as a highly approachable member of the household — such as a pet? Careful consideration of some problems can sometimes reveal powerful reasons for presentation to be treated very differently from content.

When it is important to introduce a high element of fashion into design solutions, then particular attention has to be paid to presentation and, perhaps, designers will proceed from presentation to content. In fields where the rate of innovation is low and differences in content of competitive products negligible, then, too, it is presentation that can introduce marketable differences amongst the products.

By contrast, when developing brand-new product concepts, the design of content is generally the top concern. Attention focuses on what the new product is and what it does; often presentation is neglected, although the way the product is introduced to the market can have a critical effect on its success. As innovations gain acceptance, become established in markets and rival products are launched, attention gradually shifts to presentation. Typically towards the end of a product life-cycle, design attention will focus on presentation.

It is often suggested that the wider and more competitive the market, the greater influence presentation has on the decision to purchase. It follows that where markets are confined to a handful of buyers — say, in a branch of heavy industry — the less presentation will influence decisions. Reality is not quite that simple. Firstly, the influence presentation has in purchase decisions has less to do with the breadth of a market than with the immediacy and accuracy of feedback on the performance of content. For example, where the performance of a machine can be evaluated quickly and accurately, the styling of the machine casing may have little impact on the choice for or against that machine. However, where performance cannot be gauged with that kind of certainty, then presentation is more likely to be taken into account. Furthermore, where the purchaser is also the user, the more potent is 'on-product' presentation likely to be. Thus when manufacturers in heavy engineering say that presentation is not important in their products, what they usually mean is that visual design-influenced presentation is not important. A bribe paid into a Swiss numbered account for the benefit of a buyer or 'agent' is no less an element of 'presentation' than a product casing which is designed along enlightened ergonomic principles. In the first case, the benefit is for the buyer and is not designed into the product; in the second, the benefit accrues principally to the user and derives from on-product presentation. Unfortunately, it would seem that without exploring the options, too many managers place their faith in non-design, off-product inducements to purchase when presenting their products to markets. As yet, design does not hold an important place in the marketing mix.

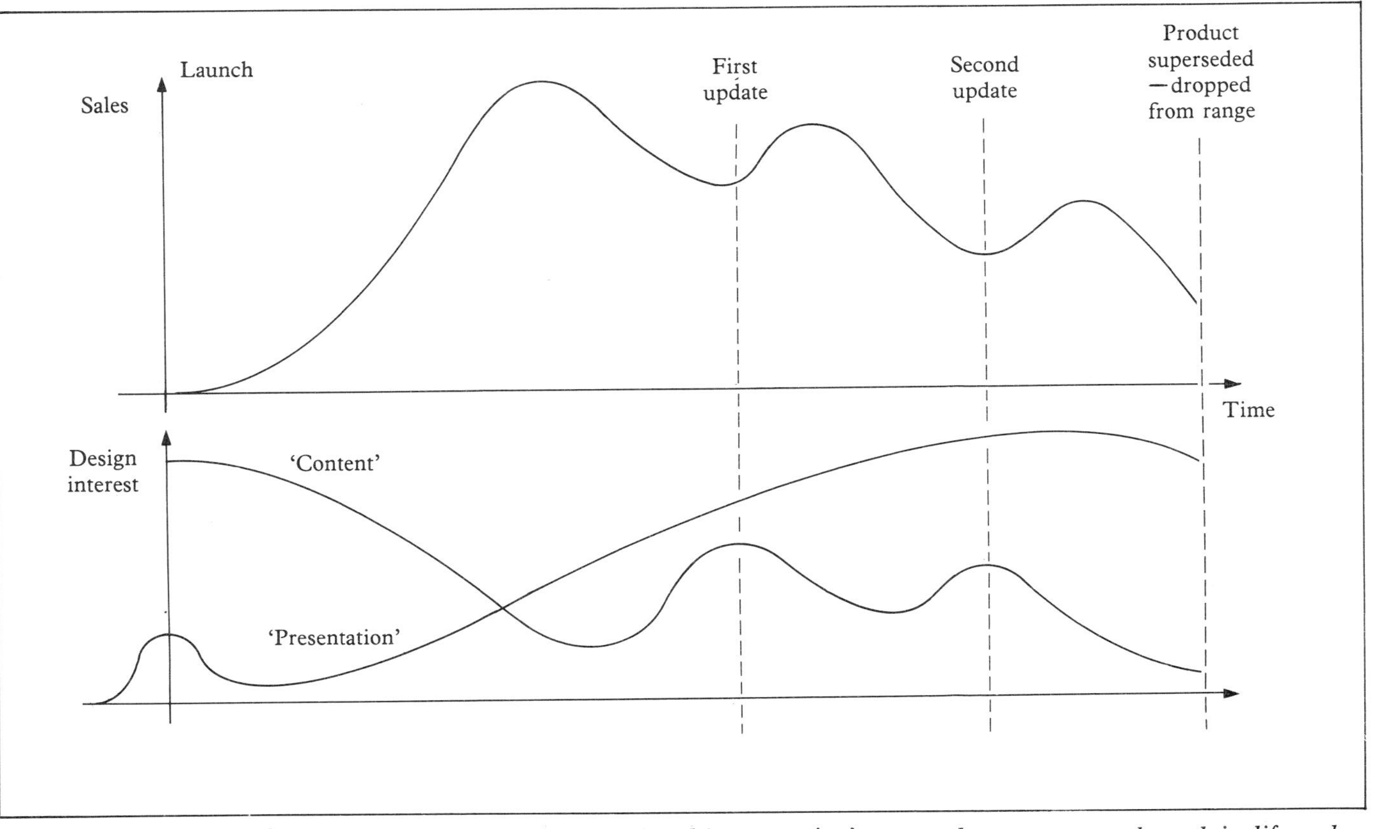

Figure 1.3 *How design attention shifts between 'content' and 'presentation' as a product progresses through its life-cycle*

Again, there may be ambiguity in interpretation, for at what point should the boundary between content and presentation be fixed? With certain types of design solution the distinction between content and presentation may be clear-cut; with others quite unclear. Indeed in some solutions the design of the presentation is little more than the design of the content — as in the case of machines where, safety regulations permitting, the mechanism is proudly exposed for all to see. Differentiating between presentation and content can be difficult enough when presentation is confined to visual design. However, should presentation be taken to encompass the wider interface between consumer/user and output, the complications mount.

Ultimately, the 'correctness' of any design orientation can only be assessed against the circumstances surrounding particular design exercises.

Design - an applied specialism

Design is seemingly unfamiliar territory to managers, despite the fact that all outputs produced by industrial and commercial organisations are 'designed'. A major problem with any discussion on design is that the term has yet to be defined in a manner which makes sense to managers. Dictionary definitions are often unhelpful, though most design spokesmen quote them.

Design is an applied specialism. The design process should always be directed towards a purpose *outside* of the designer. Thus, design for design's sake is, in fact, 'art'. The fundamental difference between designers and artists is that the latter create almost entirely for their own satisfaction: the approval of others is a plus, not an essential. The designer, on the other hand, must create for the satisfaction of others — and their approval is central to his being able to practise his profession. Without approval of the commissioning agent and consumer, the designer is rarely able to implement his designs. And the designer who does not see his designs implemented constitutes a waste.

2 Problems, design problems, and different types of design project

Consider the following problems:

1 A manufacturer wishes to help smaller retailers adapt quickly to the changes in the DIY market which will, in turn, lead to increased sales of his products. What should he do?
2 A British importer of continental products intends to market a new range of quality French tableware. Is mail-order an appropriate and profitable way to market these products? If so, what image should be adopted for the launch?
3 An international educational institution based in Europe seeks to promote itself more professionally throughout the world. What messages should it communicate about itself and what means should be used for this communication process?
4 A Citizens' Advice Bureau wants to help its staff diagnose the needs of clients in order to advise them more effectively on the benefits to which they are entitled. What form should this diagnostic tool take, given that the range of family and other benefits is constantly changing and that advisers often operate under significant pressures and constraints?
5 A manufacturer of plastic products forecasts that he will have spare capacity in one of his plants over an extended period. This spare capacity provides an opportunity to explore a new market and perhaps to raise production standards in the process. But which new markets and products might the manufacturer consider?
6 A major advertising agency operating from old and inefficient premises suffers from poor inter-office communication. This, coupled with a depressing visual environment, has led to low staff morale, resulting in high staff turnover and a poor image in the recruitment market. How might the interior of these premises be redesigned to help overcome operating inefficiencies, raise morale, and improve the agency's image in the industry?

These problems encompass a broad range of operating needs both in industry and elsewhere commonly handled by consultancy practices. What may come as a surprise, however, is that in all these instances the organisations concerned approached design groups for advice, not firms of business consultants. Clearly these problems, and hundreds like them that designers tackle every day, are far removed from the design of letterheads, product casings, the redecoration of offices, and so on — the 'cosmetic packaging' which many managers perceive to be the designer's particular province. It would seem that the spectrum of design project types — or rather the types of project designers are called upon to tackle — extends from basic, confined design exercises to the solutions of complex, wide-ranging business problems. If the spectrum of design projects is so confusingly broad, how can they be grouped sensibly in order to examine the management implications?

Two-fold problems, and a common sequence of stages

All design projects are, in the final analysis, unique. Yet it is quite untrue to suggest that no two projects are alike. There are undoubtedly differences in detail between one project and another but, more often than not, the similarities will outweigh the differences.

The principal similarity among design projects is that they all go through a common sequence of stages. A design project is normally set up because there is a perceived problem or need. The project problem is diagnosed and a brief formulated on the basis of which a solution is conceived. When the concept of solution is approved, it is designed in detail, implemented and put to use, as a result of which an impact is generated.

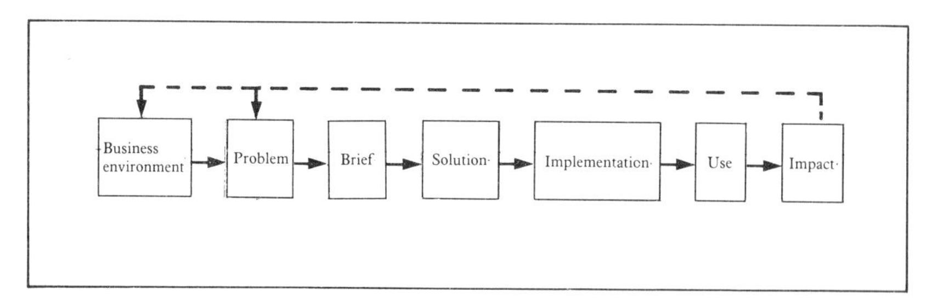

Figure 2.1 *Typical sequence of stages through which all design projects progress*

These stages can be set out in greater detail to provide an outline of the events which occur during particular stages, together with an indication of the respective contributions of client and designer to the progress of the project and the outcome.

Again, design projects of a category tend to be broadly similar. Thus, for example, projects relating to the design of interiors for offices will normally follow similar sequences of events and involve consideration of similar factors. Typical sequences of events which occur during projects relating to the design of brochures, signage

Table 2.1: Typical progress of design project, with areas of client and designer responsibility specified

Client responsibility	*Designer responsibility*
Pre-project	
Awareness of problem/need Analysis of problem/need Definition of problem/need	
Further analysis of problem with specific reference to how design skills might help	
Formulation of design brief and specification of scope of project; covering objective of project, work to be done, budget and deadline for implementation of solution	
Selection of designer	
Briefing the designer	Preliminary analysis of client problem and its background, leading either to agreement of brief, or its development/amendment. (This might take the form of a paid feasibility study)
Agree amendments to, or development of, design brief	
Establish design programme (work content and expertise, broken down into stages, with deadlines, budgets and required outputs of each stage)	Establish design programme
The Project	
Organisation of co-operative/support activity	Organisation of marketing and implementation of solution
Monitoring and supervision of project's progress	Analysis of client problem in greater detail to provide foundation for design work. In certain instances, this might involve complex surveys
Preparation for marketing of prospective design solution	Devise concept of design solution
Establishment of criteria by which proposed solution might be assessed	Interpret this concept
Assessment of solution proposed and the implications of implementing it (perhaps the solution is tested)	Formulate a specific design solution
Acceptance/rejection of solution, or request for amendments	Presentation of design solution
	Contribution to the introduction and implementation of solution. A contribution may be agreed to the marketing of the solution
Post-project	
Management of impact (i.e. monitoring and response to the reception of solution)	Development of impact
Evaluation of outcome of project	Evaluation of outcome of project

systems, exhibitions, and so on, can also be mapped out, indicating progress from the analyses of problems through to the implementation of solutions and beyond (see Appendix B).

It could be said that different categories of project tend to be different in kind. Thus a graphic design project may be said to be of a different type to an interior design project; a product design project will be different from an exhibition design project. The differences will be seen in such things as the subject matter, the materials and processes involved, the regulations to be conformed with, the specialists to be consulted, and so on. However real these differences may appear to the designer and client, they are not the fundamental factors which differentiate one project type from another from the point of view of project management. As in other areas of management, we have to look beyond the category of problems which underlie design projects in order to discover these factors.

First, it is important to reiterate that design projects involve twofold problems. Even the simplest are rarely, if ever, just about design objectives: there are underlying 'business' objectives which have to be achieved through the design solutions. Although design may be considered by many as a purely visual discipline, the contribution of design is actually judged on the impact it creates: for example, a product brochure is not designed merely because a client wants a brochure – at least, that ought not to be the case. What the client probably needs is a practical item of information which helps him answer queries, an item of promotion which makes a target audience well disposed towards his product and company, and so on. Thus, even though the client may order and accept that brochure purely on visual terms, he will assess his investment by its performance along these other 'business' dimensions.

Factors which determine the nature of design projects

The type of design project is determined partly by the nature of the underlying problem and partly by the way the project is handled.

Several factors have to be considered when determining the nature of a problem. Broadly speaking, design projects can either be concerned with problems which are narrowly defined and confined to design considerations, or they can be concerned with open-ended, wider-ranging design problems which involve critical business considerations. Again, they may be concerned with problems which occur at the periphery of the company's operations or with problems central to the very essence of the company.

Some problems recur periodically or alternatively may be anticipated; others are unique and occur quite unexpectedly. Some problems take a long time to solve while others can be solved almost as soon as they are perceived. Some problems cost a great deal to solve, however short the time spent devising their solutions; others require minimal resources. Some problems deteriorate rapidly if not tackled immediately, while others can be left unheeded for years without major change.

The solutions to certain problems will represent broad and long-term corporate commitments: their effects may not be easily amended or reversed, so the cost of making a wrong decision is significant. With other problems, the implications of decisions will be restricted in scope and limited to the short term: the time and cost involved in reversing such decisions may be insignificant.

Problems may be highly specialised in nature and require solutions close to the 'state of the art'. Alternatively, they may be of a more general nature which can be solved by means of well-tried solutions. Whether of a specialised or a general nature, problems may represent familiar or unfamiliar work to managers and others in the company. Obviously problems of a specialised nature will usually require the skills of specialists, whereas the more common types of problem may be handled equally effectively by general practitioners.

All these factors affect the work required in solving problems, hence the scope of design projects.

Table 2.2 Factors affecting the nature of design problems

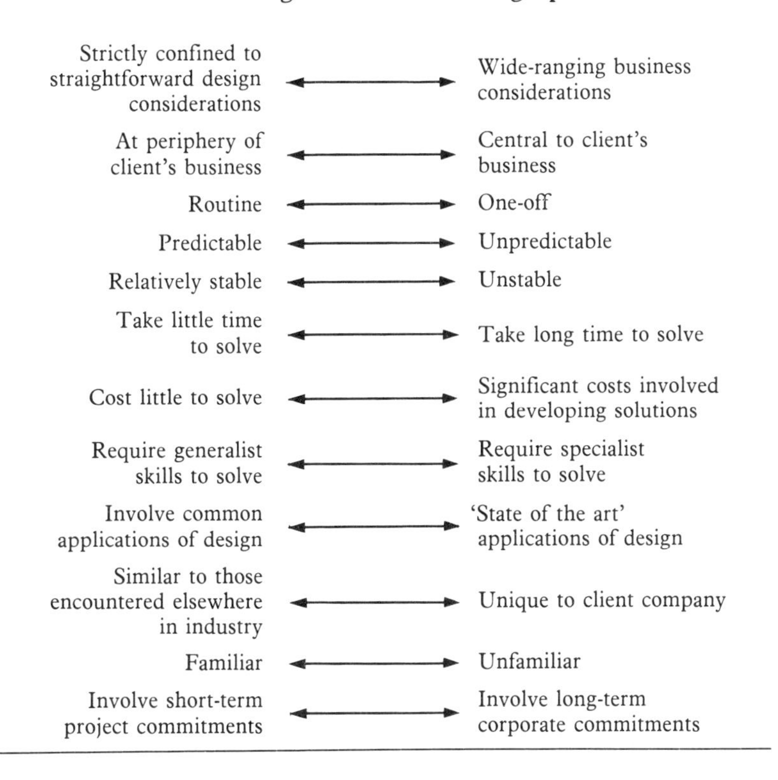

Strictly confined to straightforward design considerations	⟷	Wide-ranging business considerations
At periphery of client's business	⟷	Central to client's business
Routine	⟷	One-off
Predictable	⟷	Unpredictable
Relatively stable	⟷	Unstable
Take little time to solve	⟷	Take long time to solve
Cost little to solve	⟷	Significant costs involved in developing solutions
Require generalist skills to solve	⟷	Require specialist skills to solve
Involve common applications of design	⟷	'State of the art' applications of design
Similar to those encountered elsewhere in industry	⟷	Unique to client company
Familiar	⟷	Unfamiliar
Involve short-term project commitments	⟷	Involve long-term corporate commitments

Client awareness of problems and the scope of design projects

The client's knowledge of his problem normally influences the way he structures it and how he thinks it should be tackled. In turn, both these factors influence the scope of the project and the extent of the designer's involvement. So how much does the client know of his business problem and objectives?

At the one extreme, there are projects where the client has no idea exactly what his business problem is. He is aware that a problem exists and may be able to indicate which parts of his organisation ought to be investigated. But, perhaps because the client is unqualified or has no time, the business problem is neither diagnosed nor defined in a brief.

More often than not, these projects are primarily business consultancy assignments: no design work may derive from many of them. The client really needs a business consultant, but nevertheless entrusts the work to a design consultant who has to use his complete range of problem-solving skills in advising his client.

If the client does not allow the designer the opportunity to analyse the problem in depth, then the project may turn into a superficial exercise with neither client nor designer having any clear idea of the underlying problem. And it is not all that uncommon for projects to proceed with the client and designer actively disagreeing in their diagnoses of the problem.

At the other extreme, there are those projects where clients are fully aware of their business problems. With these types of project, the client endeavours to interest the designer in a defined business problem, expecting him to respond with a concept of a design solution, together with recommendations as to how the concept is to be interpreted into a specific solution which is then to be implemented. This can be termed a fully-fledged design assignment. An example of this type of project may be the search for new product concepts.

In describing the problem, the client may confine himself entirely to business considerations; he does not translate the business problem into a corresponding design problem. This may be because he is unable to formulate the design problem or because he deliberately refrains from so doing in order that the designer may have greater freedom to devise an effective solution according to his own perception of the design problem. However, a full design project brief would incorporate definitions of both the business and design problems.

Alternatively, in briefing the designer, the client may define both the design and business problems *and* put forward a suggestion for the solution.

Clearly, the client's awareness of the possible solutions to his problem and his views on what types of solution would be acceptable also influence the scope of design projects and the potential

Table 2.3 Spectrum of client awareness of problem

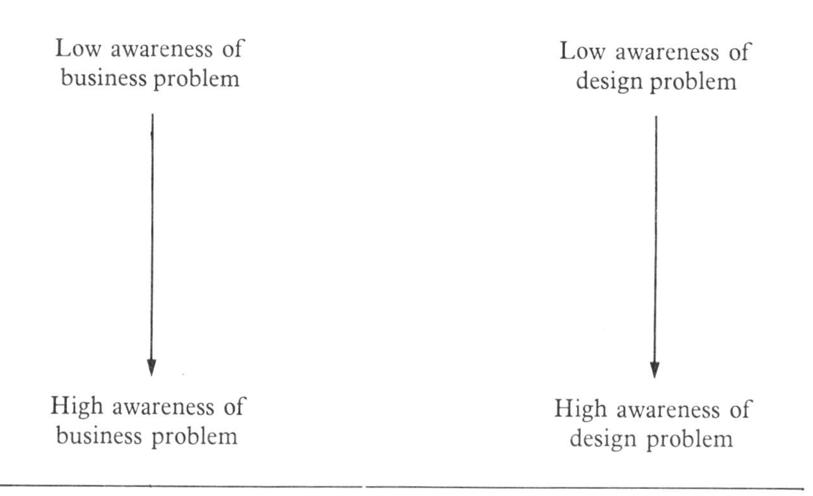

Table 2.4 Spectrum of client awareness of solution

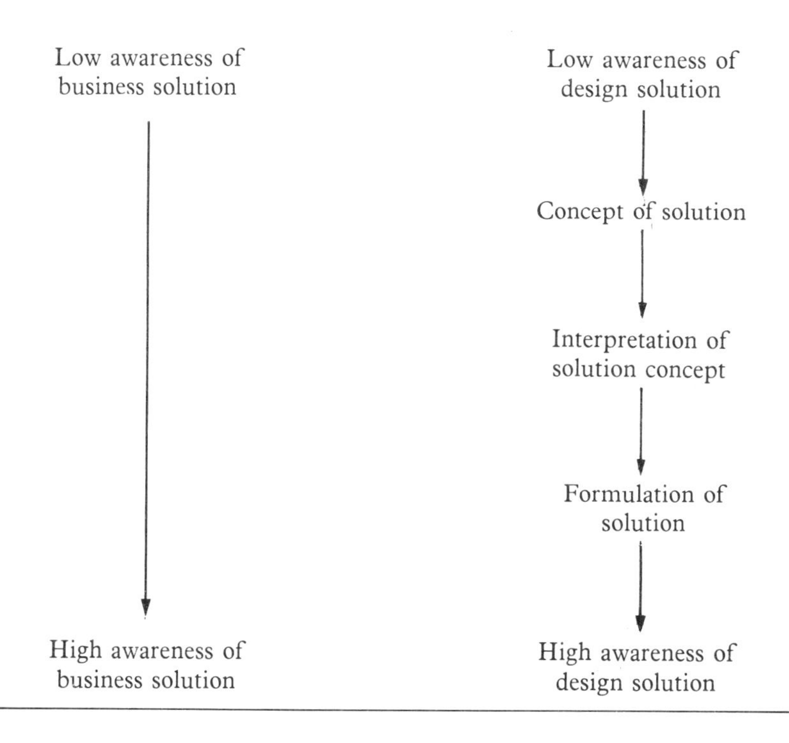

involvement of the designer. At one extreme there are clients who have no idea as to the ranges of solutions available to them. Though they define their business problems in detail, they are unable to visualise appropriate design solutions. Normally these clients will also be unable to translate business problems into design problems.

At the other extreme, there are projects in which the clients have very clear ideas as to the solutions they seek. These can turn out to be very restricted design exercises: knowing exactly what they want, the clients use the designer merely as 'a pair of hands' in order to transform their ideas into reality. From the designer's point of view, such projects may involve little more than good draughtsmanship. With so little demand for creativity, an 'independent mind' could well be at a disadvantage when dealing with such clients. However, a word of warning: designers should beware of those clients who are highly articulate on the required design solution, but remain quite ignorant of their business problem.

Between these extremes, there are degrees of client awareness of possible solutions. In some cases, the client will be fully aware of the range of solutions available to him but will feel unsure as to which represents the most effective option. In other cases, the client may have actually worked out the concept of a solution but then proceeds no further: the designer is instructed to interpret the concept and to formulate a solution. When a designer is briefed to update or restyle

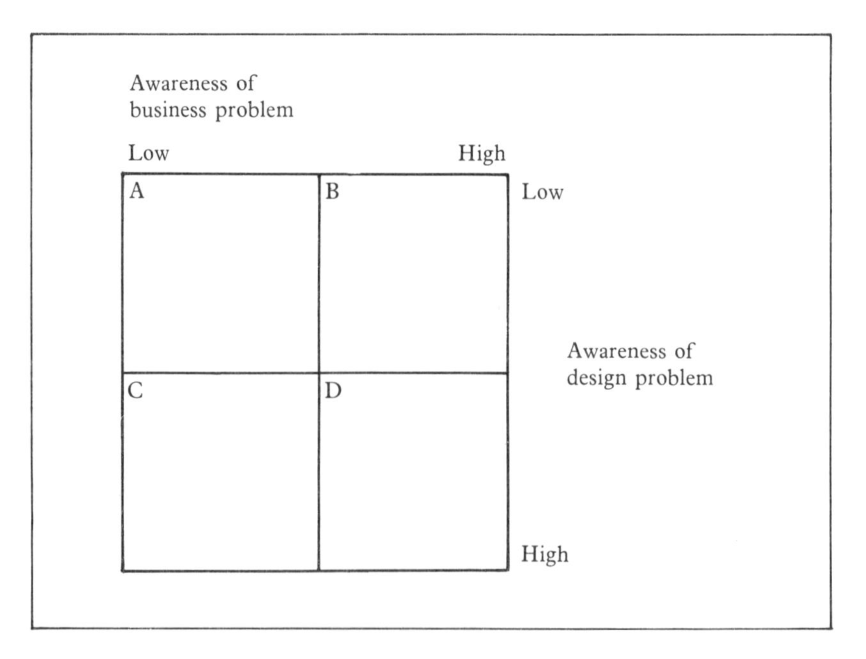

Figure 2.2 *Matrix of client awareness of business and design problems*

an existing product, or to design a minor extension to an established product range, he will normally have this kind of assignment on his hands.

A matrix can be constructed to depict the client's awareness of the business and design problems, as Figure 2.2. In A, the client has little or no idea of either his business or design problems. The designer has a full business/design consultancy assignment on his hands; his point of introduction to the project is at the diagnosis of the 'problem' stage (see Figure 2.1), and thus should have maximum influence on his client. This influence decreases as we move towards D, where the client has particularly clear awareness of his business problem and also knows what design solution he needs. It is in these circumstances that the designer acts principally as 'a pair of hands' implementing the client's requirements. His point of introduction to the project is at (or just before) the 'solution' stage, and thus his influence tends to be minimal.

Normally designers are expected to implement the design proposals they put forward to their clients. However, there are occasions when the client decides to implement the proposal himself without any involvement by the designer. These examples also represent a different type of design project — especially where project evaluation is concerned.

This matrix hints at further factors which help determine how design projects are handled, and hence their scope.

Complexity of problems, range of contacts, and discretion allowed

The complexity of the problem as set out in the brief will give some indication of the range of contacts necessary to achieve a satisfactory outcome. With a straightforward, confined design exercise, the project could be executed by an individual working on his own with no need for any contact other than that with the project manager — his 'client'. As the complexity of the project increases, the need for contacts will increase: the designer may seek information and/or advice from a limited number of contacts inside and outside the client organisation. If he is handling a complex problem on his own, the range of contacts may be extensive.

Teams set up to handle design projects have similar needs to maintain contacts. The members of these teams may be appointed on a full-time, or on a part-time basis. There may be a core of full-time members with a number of peripheral members, specialists perhaps, who are brought in only at specific stages. The members of teams may be drawn from one discipline or from several disciplines. Normally, multi-disciplinary teams will have wide ranges of established contacts. Nevertheless, designers or design teams tackling unfamiliar problems may need to discover many new sources of information and advice which adds to the specified work load.

Apart from requiring more extensive, multi-disciplinary teams and contact networks, the more complex type of design project tends also to require a wider span of seniority among team members. Typically, the more straightforward the exercise, the lower down the client organisation will responsibility for the project be vested. The greater the variety and the more complicated the work involved, the greater will probably be the range of skills used and the more the levels of responsibility involved. The more important the project to the client organisation, the more senior will be the executive given responsibility for it.

Table 2.5 Factors affecting the way design projects are handled

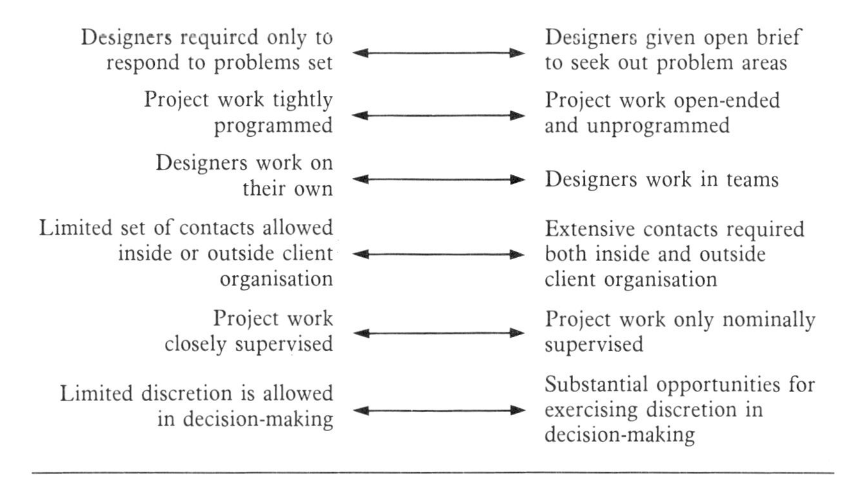

Designers required only to respond to problems set	↔	Designers given open brief to seek out problem areas
Project work tightly programmed	↔	Project work open-ended and unprogrammed
Designers work on their own	↔	Designers work in teams
Limited set of contacts allowed inside or outside client organisation	↔	Extensive contacts required both inside and outside client organisation
Project work closely supervised	↔	Project work only nominally supervised
Limited discretion is allowed in decision-making	↔	Substantial opportunities for exercising discretion in decision-making

Furthermore, how much discretion is allowed in the execution of the work on design projects? For the amount of discretion involved is a major factor to consider when determining the scope of a project. In certain cases, designers may be required to respond only to problems set them, whereas others may be given an open brief to seek out problem areas. For confined design exercises, the work may be prescribed and fully programmed. With more extensive projects, design and operational guidelines (or other 'givens') may limit the amount of discretion which can be exercised. With projects upon which no guidelines are imposed, the project manager (if not his team members) will have substantial opportunities to exercise his discretion: the project thus defined can be said to be 'non-programmed'. Non-programmed projects may usually be considered to require particularly close supervision, but fully programmed projects receive only nominal supervision. Non-programmed projects which are also only nominally supervised offer the design project manager and his team the greatest potential for exercising discretion.

Most of this type of project will be located in square A of the matrix; fully programmed projects which are closely supervised will be typically represented in square D.

Audits of corporate design activities

Labelling design projects through this kind of analysis produces several practical benefits. Certainly it helps managers to understand better the characteristics of such projects, and to relate them more closely to the operations of their companies.

More significantly, it enables managers to draw up audits of design activities within their companies which help determine the range and complexity of such activities, while the nature and volume of problems tackled should give indications of the importance of these activities — though they may not be judged similarly by all departments of a company.

In preparing such audits, questions such as these should be answered:

1 What design problems does the company have at present?
2 In which of the company's operations do these problems occur?
3 Are any of these problems grouped together from the technical or administrative points of view?
4 Is this the normal pattern of problems; and
5 Will this pattern continue for the foreseeable future?

Furthermore, the audits should enable managers to determine who gets involved in design activities. Specifically:

1 The number and positions of staff members.
2 At what level responsibility for design projects is vested.
3 The independent specialists and suppliers used, if any.
4 The range of skills they possess.
5 Whether they work predominantly as individuals, in single-disciplinary or multi-disciplinary teams.
6 The range of contacts they are required to maintain during the course of design projects; and
7 The patterns of interaction between these various people.

Finally, the audit should provide a record of how this activity is managed:

1 Are design projects administered on a centralised or decentralised basis?
2 Do any common operational procedures exist for undertaking design work?
3 How are corporate design requirements analysed?
4 Have any design guidelines been formulated?
5 How are design investments monitored and evaluated? and so on.

With a clearer picture of design activities and requirements within their organisations, and a greater understanding of the nature of design projects, managers should be in a position to manage design projects more effectively, *and with greater confidence.* In essence, audits of corporate design activities offer firmer foundations on which to plan ahead and organise investments in design. As such they are an invaluable management tool.

3 The pre-project phase in design projects: work leading up to a brief

It is almost as useless to produce a brilliant solution to the wrong problem as it is to produce an ineffective solution to the real problem. Consequently, the effort put into problem diagnosis and the care with which problems are defined are matters of crucial importance: correct diagnosis and accurate definitions of problems are the twin foundations for successful design projects. Yet it is precisely during this ground-breaking, 'pre-project' phase of such projects that a lack of resources and commitment is most often apparent. Why is this so? Reasons can be found both in management perceptions of design and in the importance attached to the contribution design can make to profitability. Reasons can also be found in the operating circumstances in which many design projects are set up.

The myth of the superficial exercise

The myth that design projects are, by and large, superficial exercises persists in the minds of managers — particularly those who have no experience of working alongside professional designers. Thus, managers will suggest that as design projects are largely about 'cosmetics' and involve highly subjective decisions, they deserve no more than superficial attention. This applies especially before funds are allocated and projects are set up formally — which is precisely the time when most problems are diagnosed.

Other managers believe that design problems are relatively straightforward to define. To many, such problems tend to be 'glaringly obvious', perhaps because they derive from 'familiar' business/operational requirements. The necessity to translate carefully documented and diagnosed business problems into workable design problems is not always grasped.

Still others will say that designers are not able to assimilate detailed analyses of business matters, and so only 'basic facts' are gathered and communicated. To reinforce this practice, it is also suggested that providing too much detail and organising rigorously puts a constraint upon creativity.

It is true that design problems can arise with alarming rapidity and they sometimes require immediate solution. In such instances it could be claimed with justification that there is little opportunity for analysis in depth; managers and designers may have to react instinctively when diagnosing the problem and generating a solution. Yet it is fair to state that managers rarely take design into account when planning: the designer's skills tend to be used as operational tools often on an immediate, 'fire-fighting' basis. Consequently, design problems are tackled last of all or go unheeded until the crisis develops; time is then short and very quick reactions are necessary.

Many design projects are set up on the basis of ill-defined problems because either designer or client – and sometimes both parties – are too anxious to get the project confirmed and under way. A project with a first stage which involves researching the problem in order to formulate an effective brief tends to be considered a luxury, if not an admission that the client has failed to do his homework. Therefore a preliminary outline of the problem will be included in the project proposal and, as a result of operating and/or political pressures, will remain unaltered as the project progresses and it thus becomes the only available 'definition' of the problem.

Finally, some managers contend that it is really the designer's job to diagnose design problems, after being given an indication of the client requirements. A high proportion of design projects do proceed on the basis of proposals submitted by designers to their clients. These proposals will invariably contain the only record of the problem. Some designers are as thorough as can reasonably be expected, taking time to examine the potential client's particular circumstances before defining the problem; they may not be paid for this involvement and the potential client may not cooperate in the exercise, considering this work to be part of the 'job getting' effort by the designer: 'let's see what they come up with' is a common attitude. If the designer has no knowledge or experience of the potential client's organisation or the industry in question, and if the potential client does not participate in the diagnosis of the problem, then the chances of ending up with an accurate definition of the problem are slimmer than should otherwise be the case. Inevitably, there are designers who do little more than relay back to their client the main points discussed during an initial meeting; in some cases, the points will be repackaged to give the semblance of an authoritative definition.

Clearly it is not always possible to diagnose a problem correctly or to define it accurately: there is just so much which can be known about a problem before work starts on solving it. Indeed, it is sometimes argued that no problem definition is complete without some reference to possible solutions, and that referring to one set of solutions as opposed to another effectively changes the definition of problems. It is a fine point whether problem identification and problem solving are two quite separate activities which can be undertaken, each divorced from the other. There is certainly a class of problems for which the definitions at the same time constitute specific, and sometimes unique, prescriptions of the form the solutions will take. There is also a class of problems which does not consist strictly of problems or needs, so much as of 'wants': that is, a designed output is clearly visualised without there being a problem other than the challenge of turning vision into reality. However, not all problems that are defined in terms of solutions are of this class. Many clients fall into the trap of jumping to conclusions about necessary, or even possible, solutions without analysing their problems systematically. Consequently, their 'definitions' tend to give more insights into their expectations of acceptable solutions than into the nature of the problem.

Clearly, too, certain problems defy concrete definition. We become aware of them but are unable to pinpoint their causes, or anticipate their outcomes. In these and many other instances several equally plausible structures might be ascribed to each problem: there appear to be no obviously 'right' diagnoses or definitions, and equally there appear to be no obviously 'right' solutions. Few design problems have unique definitions or solutions.

Finally, it should be borne in mind that there is a point beyond which additional information on a problem is unlikely to help create a more effective solution, and may even become a hindrance. Even after a project is completed, aspects of the original problem may remain unknown.

What, then, is involved in defining problems?

Problems in context

Problems do not exist in a vacuum: they derive from particular circumstances. We become aware of a problem as a result of a stimulus or a series of stimuli; and every problem occurs within a context. Both stimulus and context may affect the very nature of a problem or merely the perception of it, so that when a junior executive suggests that his company's offices are badly designed, his superiors may not accept a problem exists at all, even though the arguments put forward are both detailed and sound. Yet a hint of similar concern from the managing director may be sufficient to convince his colleagues that a very serious problem exists. If workers strike because they state their machines are 'outdated and unsafe', there is a strong pull to diagnose the problem as one of 'dangerous machinery'. Yet the workers may have several grievances, of which 'dangerous machinery' is low down in priority but is the easiest to articulate and take action on. In this instance, re-equipping the factory is unlikely to have much effect on worker-management relations. However, if management acknowledge some of the high-priority grievances and endeavour to settle these, it may *still* be necessary to replace a token number of machines first as a face-saving gesture towards the workers. Thus, even if a stimulus has no effect on the diagnosis of a problem, it may still have an effect on how the 'real' problem is handled. Of course, with many problems the stimulus serves only to raise awareness and has no influence on either the analysis of the problem or the way it is tackled.

A need set against one background may well require very different handling from the same need described within an alternative environment. For example, the design of a product label may have to be altered radically for statutory and cultural reasons if the product were to be exported from one country to another. More information may need to be given on the label; the emphasis given to certain items of information may have to be changed; the information may need to appear in more than one language. Even colours may have to be altered. What appeared to be an almost identical need requiring only minor alterations is transformed into a fairly complex exercise.

Such transformations can also result from the depth to which a given background is analysed, and from the breadth of view taken. Designing a product brochure with only the consumer in mind could be treated as being no different from the problem of designing a brochure for the same product which is aimed at the retailer as well as the consumer. That is, the requirements could be considered as identical. However, a more detailed analysis of the latter problem would probably reveal differences in information requirements or differences in priorities amongst requirements. What impresses the consumer about the product, and thus influences him to purchase, may not impress the retailer sufficiently to persuade him to stock it; what retailers wish to know about the product may not always interest consumers, and so on. A narrow view of a problem may close the manager's mind to a whole spectrum of different effective solutions because it masks the potential area of manoeuvre in creating a solution: common definitions of problems will normally elicit familiar solutions. When dealing with problems perceived to be relatively unimportant or problems which are 'so familiar' that their solutions are 'obvious', originality in approach may not be appreciated.

A problem perceived within a particular set of circumstances using one form of analysis may be diagnosed quite differently if another analytical approach were adopted. Different people using the same analytical aproach may also end up with differing diagnoses of the same problem. These differences arise from the way problems are structured, hence the kind of information gathered; from the confidence with which problems are analysed, the uncertainty involved and the assumptions made; from the ease with which data are manipulated, as well as from the expectations of what might be achieved with the solutions generated.

Thus problems should be analysed against specified backgrounds and the solutions generated should be appropriate to these contexts. This statement may be tritely self-evident, yet it is not uncommon for design problems to be 'defined' and 'solved' without proper reference to context. Indeed, a fair proportion of design-problem definitions turn out not to be definitions at all but prescriptions for solutions, and many design proposals which superficially offer solutions actually generate more problems than they solve.

Furthermore, problems are often viewed as being static: once defined, they are assumed not to change — at least while work progresses towards the creation of solutions. This may be true for relatively straightforward design projects of short duration. But many design projects stretch out over several months, if not years, and deal with complex problems within rapidly changing environments. What was considered a minor shortcoming yesterday can become a pressing need tomorrow; an aspect of a problem which is given relatively low priority at one stage may turn critical if left unsolved; what is perceived as a problem today may, for all sorts of reasons, cease to be one in the future, and so on. Thus the definition of a problem ought to include a time reference during which information gathered on the nature of the problem and its context is expected to remain relevant. Once defined, problems should be tackled within a prescribed time-scale, with due consideration for changes which might occur.

The articulation of problems

Diagnosing problems correctly is only part of the battle leading to accurate problem definitions. Diagnoses of problems need to be communicated between client and designer: the definition of a problem, whether oral or written, constitutes part of this communication process, but the communication process itself may influence the definition of problems.

We have seen how the kind of information gathered and the way it is collated may vary from person to person, even though they are diagnosing the same problem. As people differ in the language they use, with regard to terms and the assumptions underlying such terms, in the extent to which they are articulate and in their sensitivity to others, the manner in which a problem is communicated will vary from one person to another. Consequently, when two people arrive at an identical diagnosis of a problem, they often define it in differing terms. Indeed so different may be the terms used that the impression gained is of different problems being diagnosed, or that the diagnoses are at variance. This does not imply that there is only one way by which a given problem can be communicated effectively, for there are obviously many ways of conveying the essence of a problem while ensuring the message is received accurately. But as the vast majority of design-project briefs are communicated verbally with no supporting written material, the pitfalls inherent in verbal briefing are of particular interest in the management of design projects.

Consider the process by which a problem is diagnosed and communicated.

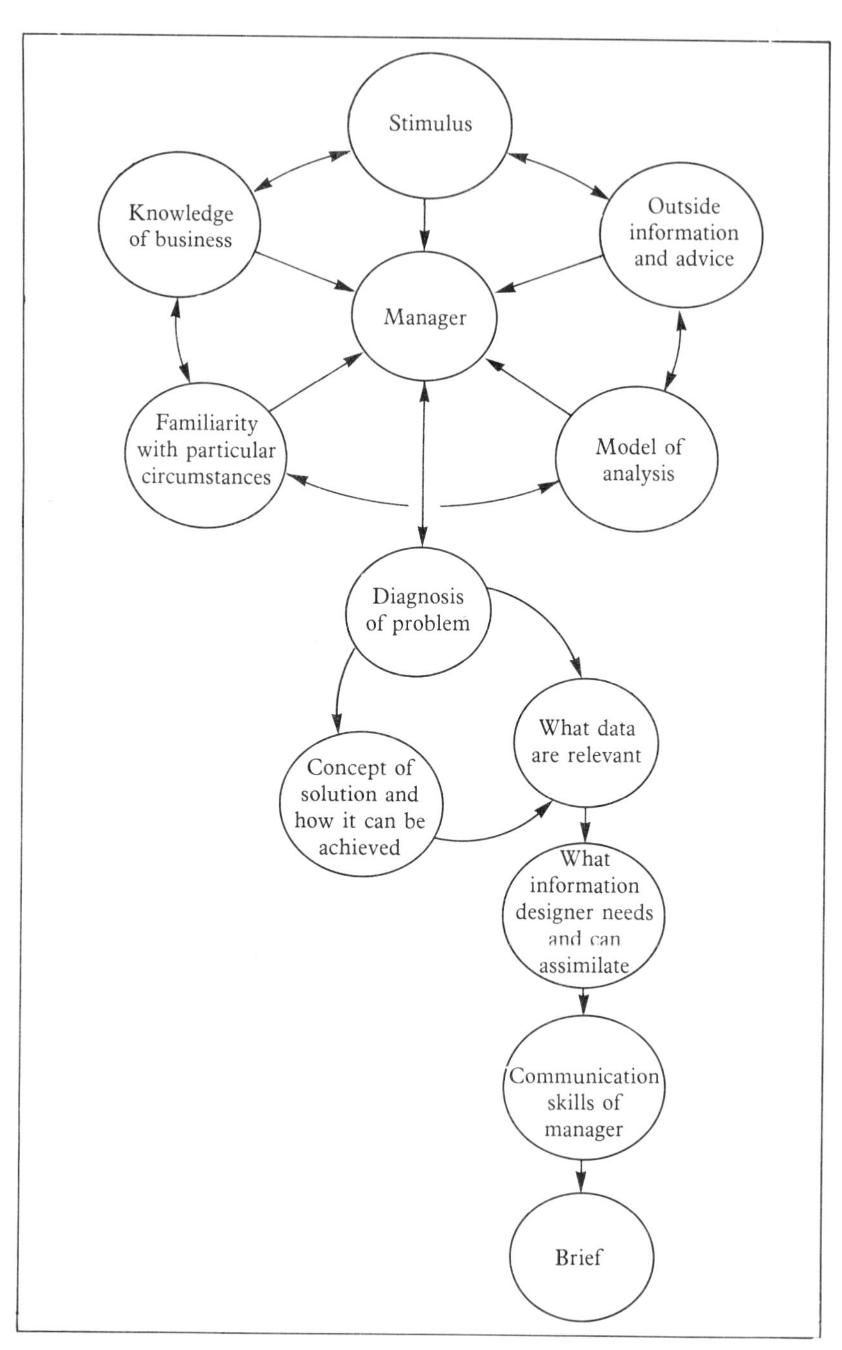

Figure 3.1 *The process by which a problem is diagnosed and communicated*

A manager becomes aware of a problem by way of a stimulus. Through his knowledge and familiarity with the particular circumstances from which the problem arises, his model of analysis, and the information and advice he seeks from within and outside his organisation, the manager diagnoses the problem. The route by which the diagnosis is reached and the diagnosis itself indicate to him what factors and data are important to the definition of the problem. The manager also gains an impression of the kind of solution (or solutions) which would be appropriate, and how such solutions might be achieved and implemented (Figure 3.1).

At this stage, attention will focus on the information requirements of a designer. The manager will interpret what he considers to be the designer's needs and will form an opinion as to the amount of data and detail that a designer can assimilate. Nevertheless, the manager may articulate the problem, using terms which are not necessarily familiar to the designer, or he may make assumptions which are not shared. In the absence of written material and penetrating discussion, this articulation of the problem becomes the project brief — a brief which might be quite inaccurate or very biased as a result of the various filtering processes the data have been subjected to. If now the designer decodes the messages he receives differently from the way they were encoded (which is highly probable), then there is clearly a further source of confusion: however clear the problem articulated by the client might be, the designer understands it differently.

The missing link: the 'operating' problem

Every problem is affected by a variety of factors which derive from its context. Some of these factors have direct and strong influences on it, others are less critical. A problem is unlikely to be solved effectively unless the first set of factors is considered and dealt with, where necessary. This constitutes the *problem management area.* The other factors go to make up the particular universe of the problem — the *problem hinterland.* Thus each problem has a problem management area and a problem hinterland, and it is the boundary conditions between problem-problem management area and problem management area-problem hinterland that define the scope of design projects (Figure 3.2).

As explained before, the way a problem is articulated can affect the received diagnosis of the problem and hence the structure ascribed to it. The way a problem is articulated may thus affect the extent of the problem management area and the problem hinterland.

Consider the following variants of a 'brief' from a multi-product company: (a) We need a new brochure to promote product Q; (b) We need a new brochure to promote product Q which will also be used as a corporate brochure. In the second case, does the client require a product brochure with a 'corporate' bit added on, or does he really

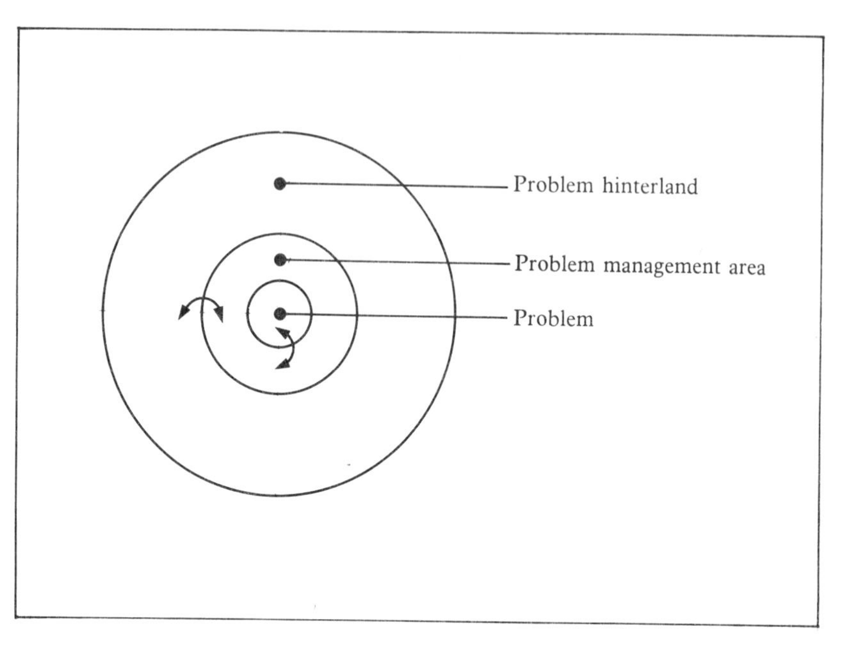

Figure 3.2 *Problems, problem management areas, and problem hinterlands*

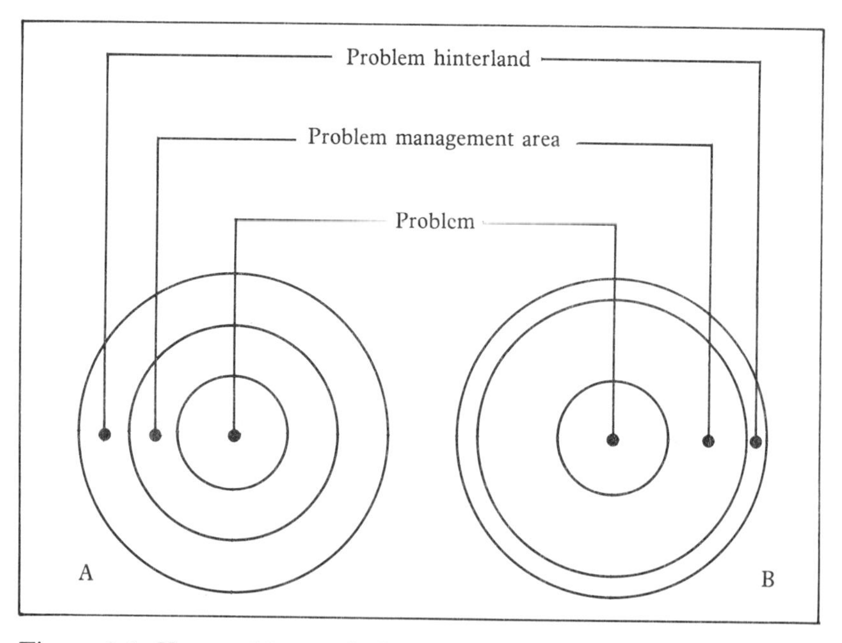

Figure 3.3 *How problem articulation affects the extent of the problem management area*

seek a corporate brochure with a 'product' bit added on? In (a), the problem clearly majors on product Q and the problem management area is accordingly restricted. In (b), the problem, though articulated in almost identical terms, involves a problem management area which extends over almost the whole of the problem hinterland.

Though the final designs produced may not look strikingly different, the scope of the second project is substantially greater than that of the first, a fact which will certainly be reflected in the amount of work undertaken before a solution is formulated. This point also illustrates how apparently minor differences in problem articulation can alter both the nature and scope of design projects, even though the definitions may be based on identical diagnoses of problems.

What if the problem has not been diagnosed or articulated correctly? One of the priorities of a designer when discussing new projects should be to ascertain whether the problem as articulated by the client is, in fact, the real problem. Furthermore, is the problem an entity, or is it an amalgam of related problems? Once more, take the example of the brochure for product Q. The client sees his sales dipping and notices that competitive products are all supported with prestige brochures. An 'obvious' conclusion is that there is a need for a new product brochure; this is the need communicated to the designer. The real problem may be quite different: perhaps the product specification has slipped vis-à-vis the competition. Alternatively, representatives from the client company may be less effective in dealing with retailers: consequently, less display space is given to product Q and it is not pushed as much as the competition. Of course, neither of these problems is likely to be solved with a brochure, prestige or otherwise. Should the designer conclude that the problem has not been properly diagnosed, he has then to convince his client that the definition needs to be altered (Figures 3.4 and 3.5).

The client may accept the designer's diagnosis and trust in his judgement, and so agree to the proposed redefinition of the problem. Alternatively the client, insisting that the original diagnosis is broadly sound, may nevertheless concede that it could benefit from amendment in one or two details — perhaps to hedge his bets. Thus he 'negotiates' an *operating problem* which incorporates some of his own diagnosis and some of the designer's (Figure 3.6).

Agreement on an operating problem is very important because it affects the work involved in the project, the apportionment of work and responsibility between designer and client, the additional specialists who may be needed, the timescale and costs of the project. These are all areas where unpleasant misunderstandings can and do occur; these are all areas which influence evaluations of design projects.

That the operating problem affects the scope of a design project can be seen in Figure 3.7. If the fit between the articulated and the real problem is poor, the problem management area is greatly expanded. In

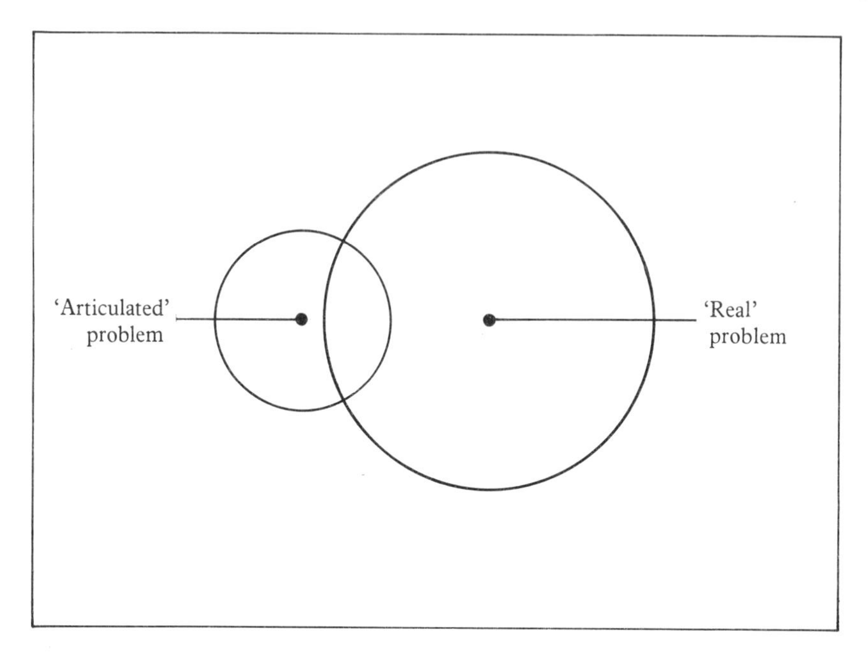

Figure 3.4 *Differentiating between the 'real' and 'articulated' problems*

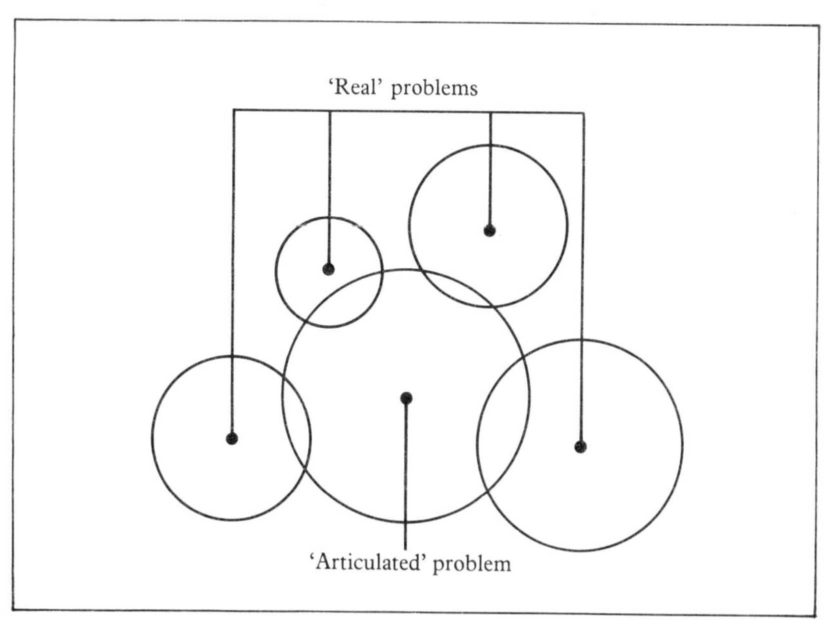

Figure 3.5 *Several 'real' problems amalgamated into one problem during the briefing process*

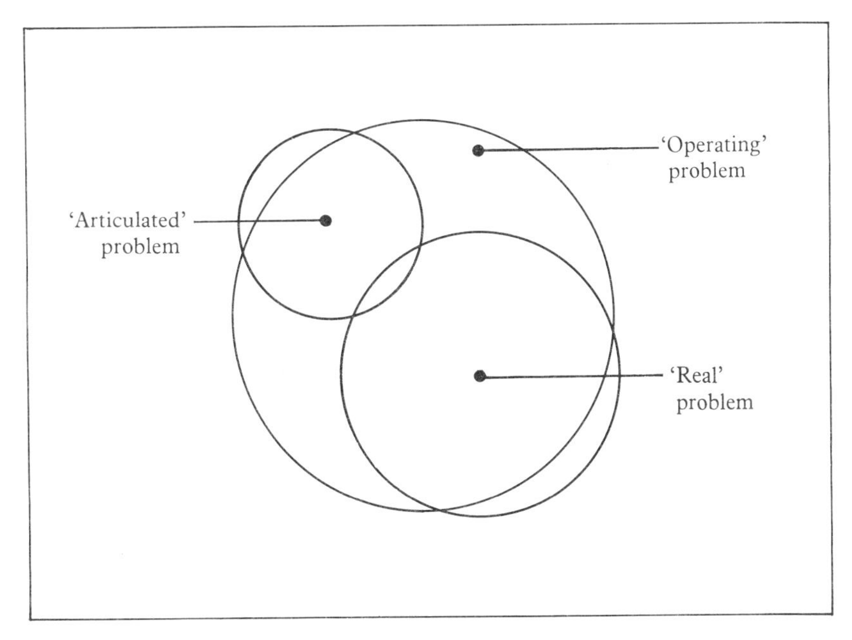

Figure 3.6 *The 'operating' problem negotiated from the 'real' and 'articulated' problems*

such cases, if the client refuses to amend the project brief, the designer may have to turn down the project – or he can proceed to devise a solution for the articulated problem in the belief that it will not solve the client's real problem. If the fit is close, the designer may choose to expand the scope of the project informally as it progresses, in the hope of being able to incorporate elements of a solution to the real problem in his design proposals. This strategy often arises out of a sense of pride in one's work and problem-solving ability. It often heralds a higher work load for the designer which the client neither acknowledges nor remunerates.

With luck, the client may agree to change the brief as the project progresses. A brief can, however, change over the span of a project, even though designer and client were in agreement over the problem at the start. There is just so much that can be known about a problem before work starts on devising a solution, and as more knowledge is gained of the problem during the course of a project, either party (or both) may come to realise that the problem is of a very different nature or that the priorities for solving it need amending. So, going back to the example of product Q, at the start of the project both designer and client agree that a new brochure is required. The designer visits a representative sample of retail outlets to see for himself how the product is bought and to seek the opinions of retailers. In the course of these visits, he notices the lack of display space allocated to the

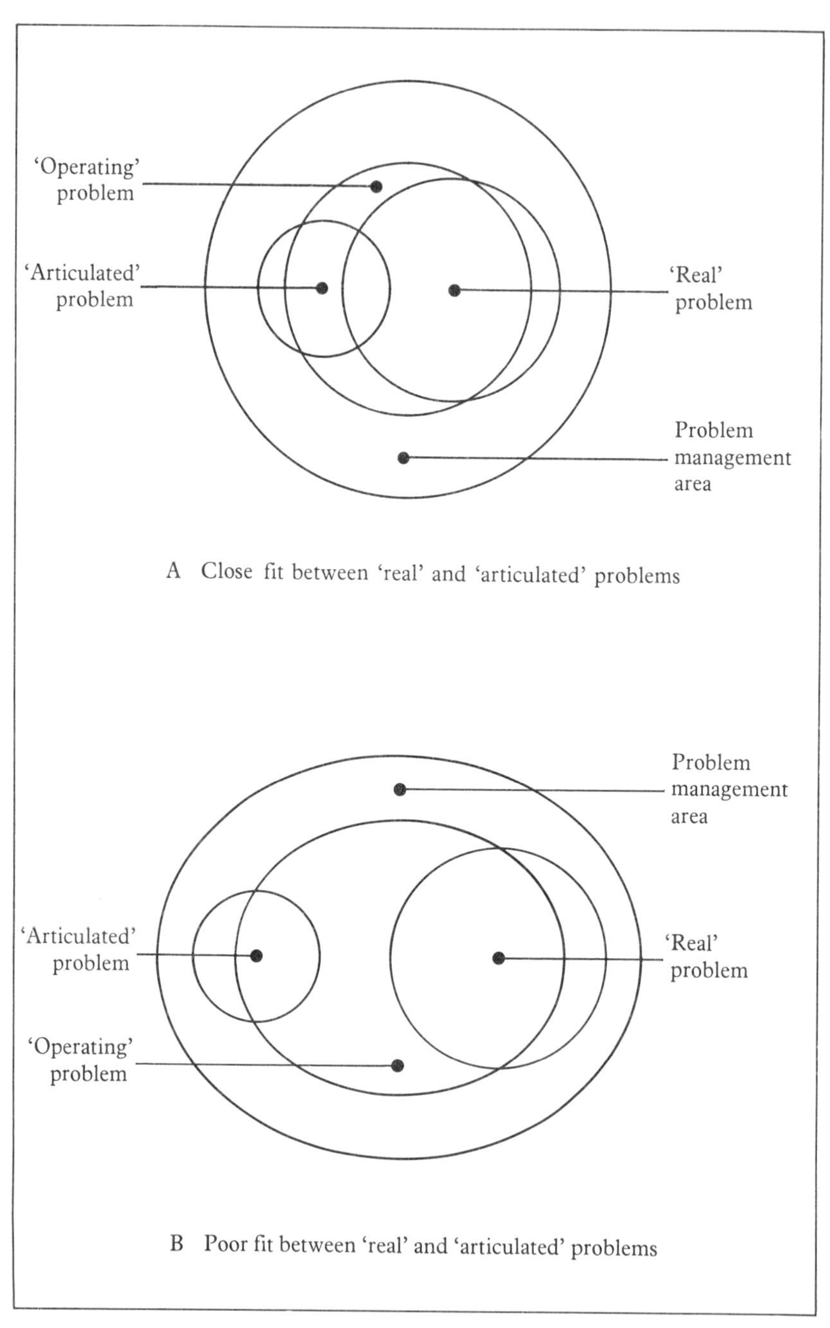

Figure 3.7 *How the operating problem affects the scope of a design project*

product; moreover, in conversation with retailers, the ineffectiveness of representatives is mentioned several times. As a result of these findings, the design requirement might be changed from a brochure to an attractive display unit which would encourage retailers to allocate more display space to the product. Alternatively, the design requirement might be changed to a presentation kit which will help representatives deal more effectively with retailers. In both instances a product brochure may also be produced, but priority is given to the other output.

Frequently it is mounting confidence which encourages the reassessment of a problem which was perhaps defined tentatively and too narrowly at the start of the project. Confidence relates not only to knowedge of the problem at hand but also to the working relationship between designer and client. Many clients will restrict the scope of design projects when they engage professional designers for the first time, or select designers they have not worked with before. Once a satisfactory working relationship has been established or a project has been completed successfully, the designers may be allowed greater scope as the project progresses or in subsequent projects with equivalent or more complex problems.

Unfortunately, the fact that a problem has been too narrowly defined may reduce the chances of developing an effective solution.

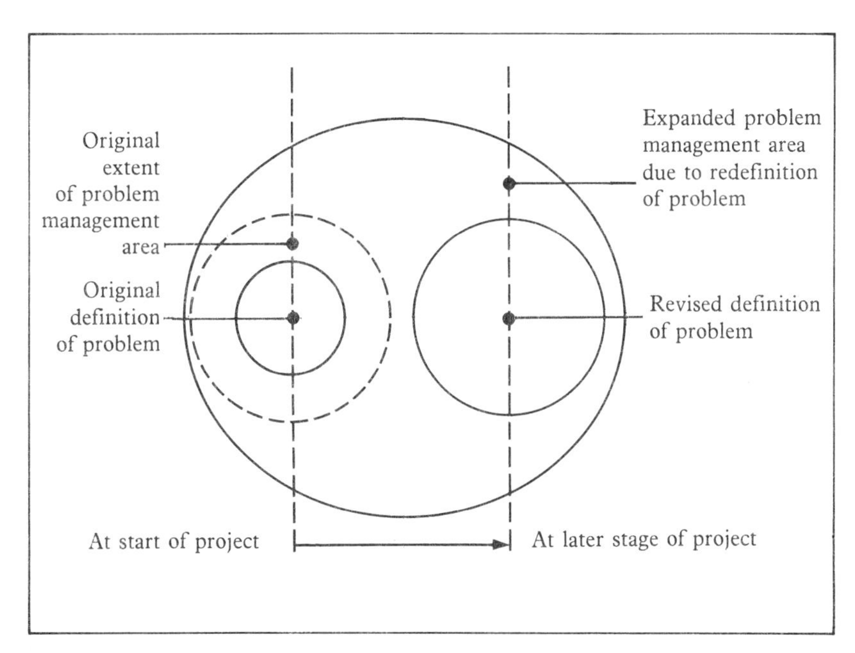

Figure 3.8 *Changes in problem definition as a design project progresses*

The client may interpret an unsatisfactory outcome as a result of the designer's incompetence.

Finally, the sequence in which the problem is tackled affects the problem management area. There is often a logical progression of work in design projects, and asking for certain results out of sequence can be very costly. Therefore, agreement on a work programme through which the operating problem will be solved is another vital component of a comprehensive project brief.

Design and the business dimension

An important point to keep in mind when preparing briefs for design projects is that design projects involve twofold problems. They are rarely (if ever) just about design objectives: there is an underlying business objective which has to be achieved through the design solution. For although design may be considered by many as a purely visual discipline, it is actually judged by the impact it creates. For example, an exhibition stand is not designed merely because the client wishes to have a framework in which to display his products: the principal objective is probably to create an attractive context through which the client can inform the largest number of potential buyers about his products as efficiently as possible. Thus, even though the client might order and accept the design of that stand on visual terms, he will assess his investment by its performance along other 'business' dimensions: the number of visitors attracted onto the stand, the number of useful contacts made and enquiries received, how well it stood up to wear and tear, and finally the comments received about it. The brief must, therefore, include statements of both the business *and* design objectives.

Sometimes the client is quite specific about his business objectives: 'We need to introduce a new product to our range because product Z will be phased out in two years' time and we need to make up a gap in sales of £100000 per annum,' he may say, or: 'I want a glossy, full-colour brochure which will treble the number of delegates at our annual conference.' Though both these 'briefs' state business objectives, there is a fundamental difference between them. The first contains a description of the problem but offers no judgement on the solution. By contrast, the second states the problem and dictates a solution: the client would seem to be fully aware of both the business and design problems. On occasion, the client will be highly articulate on the required design solution, but quite ignorant of the business problem. A matrix can be constructed of the client's awareness of his problems:

In A, the client has little or no idea of either his business or design problem: the designer has a full business/design consultancy assignment on his hands, and should have maximum influence on his

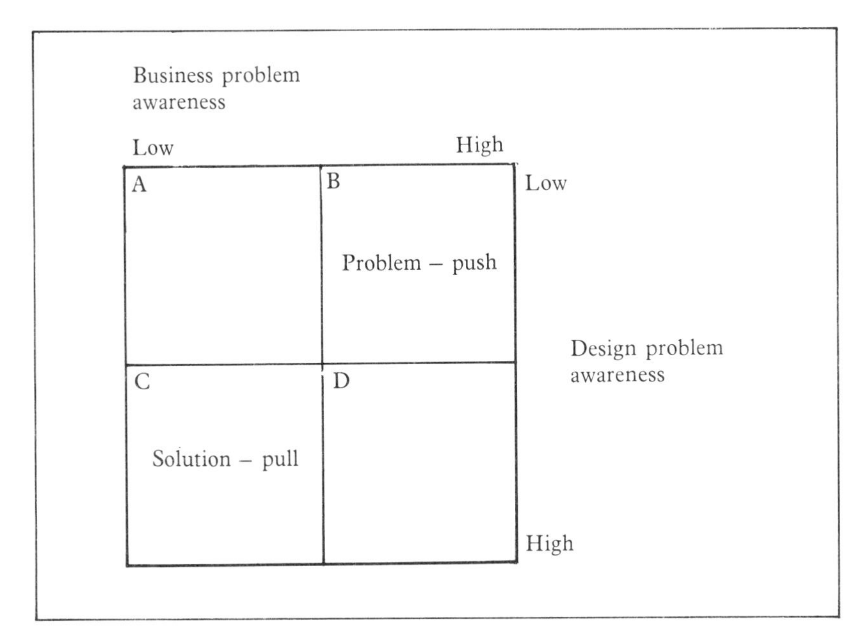

Figure 3.9 *Matrix indicating the client's awareness of his business and design problems*

client. This influence decreases as we move to D, where the client has particularly clear knowledge of his business problem and knows exactly what design solution he needs. In these circumstances, the designer acts principally as 'a pair of hands' implementing the client's requirements; his influence is minimal.

The client's awareness of his problem will affect the manner in which the brief is evolved. Interestingly, in A, it is the *designer* — not the client — who proposes the problem for solution. The formulation of the brief is entirely in his hands. In B, the client will be quite specific in describing his business problem and may go on to suggest what the design solution should *do,* but will not comment much on the *form* of the solution. The designer needs to translate the business problem into a design problem, and from there formulate a brief. He is spurred on by the articulated business problem to develop a design solution — hence the term *problem-push.* In C, the client will concentrate on the *form* of design solution he seeks, but will normally explain little as to why it is needed. The designer has to unravel the client problem by analysing *back* from the 'required' design solution articulated by the client — hence the term *solution-pull.* Frequently, the clarification of the client's business problem indicates a very different solution from that 'required'. Here again, the designer has to report back to the client; hopefully a revised brief will be drawn up.

The problem/solution balance

The designer must ascertain whether the business objectives behind the project can be achieved through a design solution. If this seems possible, are the resources made available within the project adequate to develop that type of solution? It may be that the design solution can be developed but will not generate the necessary impact unless the client initiates various 'support' activities to prepare the way.

So it is not only the problem that needs to be negotiated. The designer must ascertain quickly the acceptable forms of solution from which he will negotiate his work load and responsibilities. The client is likely to have limited knowledge of the range of design solutions available to him. Nevertheless, during briefing sessions, he may imply, suggest, or even dictate a particular solution for his problem. Frequently, the solution put forward is a 'fashionable' one, or one he has previous experience of, or even one he has seen working elsewhere — though not necessarily in similar circumstances. It may be that the cause of the problem 'indicates clearly' to him the solution required, or perhaps the method by which he analysed the problem led him to think of it.

There must always be caution when a client is particularly articulate about a specific solution. Is he 'locked into' that solution? Is his brief, therefore, biased towards this solution and/or biased against other solutions? So often a mere suggestion at a briefing becomes a strong expectation by the time solutions are presented.

There is a further danger with 'briefing with implied solution'. If the client becomes 'locked into' a particular solution, he may become blind to further analysis of the problem; he has already 'sorted it out' in his mind. The inadequate knowledge that he has of the problem could lead him to trivialise the work involved in implementing an effective solution. Hence, where appropriate, briefs for design projects should provide some formal indications of types of solutions perceived to be unacceptable, and so on.

Contents of comprehensive project briefs

Clearly the process by which briefs are evolved differs from one type of design project to another. Clearly, too, the nature of the brief varies. Nevertheless, this discussion has covered sufficient ground for a statement to be made on the contents of design project briefs.

It is worth looking again at the common sequence of stages by which design projects progress from diagnosis of problem through to the evaluation of the impact generated by the implemented solution.

The work involved in diagnosing the problem, in drawing up and negotiating a project brief, and in selecting a designer can be termed the *'pre-project'* phase. The principal objectives of the pre-project phase are to establish accurately the project problem and otherwise to lay the foundation for the creation of an effective solution.

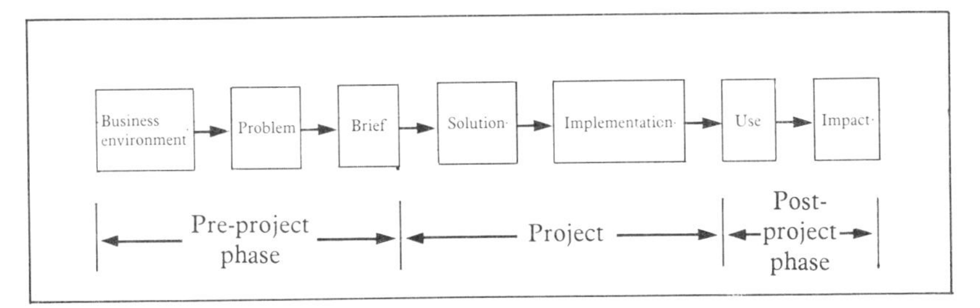

Figure 3.10 *Common sequence of stages through which design projects progress*

The brief frequently represents the point at which the efforts to diagnose and structure problems are transformed into efforts to solve those problems: briefs provide the pivotal moment at which clients and designers look *back* at the business environment which gives rise to the design project, and *forward* to the impacts which such projects should generate.

The *'project'* phase proper then extends from the brief, through the *implementation*, to the *usage of the solution* created. This phase accords closely with what most designers and their clients consider that design projects should encompass.

The *'post-project'* phase follows on from the usage of the solution to the evaluation of the impact generated and the extension of experience gained through the project into other areas of the client organisation. In essence, a *comprehensive* project brief will contain information on each of these aspects.

First, the project problem will be defined. A *business brief* describes the business problem which underlies the project, and sets it into the context of the particular circumstances of the client organisation. A *design brief* describes the design problem which has been distilled from the business problem. Both descriptions of problems are likely to contain information on the implementation of solutions (for example, in terms of the production facilities available within the client organisation), and the use to which the solutions will be put (in terms, say, of a retail strategy). The descriptions will also provide information on the impact sought through the project (a specified uplift in sales, perhaps, or a planned change in user perceptions).

In certain instances a *solutions brief* will also form part of a project brief. Through this the client may specify the solution he seeks, or else client and designer may agree on the areas of search for solutions or the types of solution which would be inappropriate.

Finally, the comprehensive project brief will incorporate a *work programme* which sets out the work involved in progressing the project through the various stages. This 'work brief' specifies the elements of work broken down into the project stages, who does what and who is responsible in the client organisation or elsewhere, resources allocated, deadlines, when and how work will be presented, and associated budgets.

Towards a shared sense of involvement

Most of the difficulties encountered during the pre-project phase in design projects are of a technical nature. The skills involved in negotiation are, of course, both technical and behavioural. However, it should always be remembered that the overriding difficulty during this phase is behavioural: it boils down to a designer and his client building up an efficient working relationship. The surest foundation to such relationships is a shared sense of the importance of the task undertaken together, and a respect for each other's professional skills.

4 Not paid to be nice guys

Consider some of the difficulties confronting a designer and his client at the start of a design project.

With a significant proportion of projects, designer and client will not have worked together before and will have little knowledge of each other's background training, breadth of experience and range of skills and what makes the other tick. It is not uncommon for the client to be totally inexperienced at dealing with professional designers since he may have, perhaps, previously entrusted design problems to suppliers such as printers and builders. Consequently, he knows neither how to set projects up, nor what timescales and budgets are reasonable.

Similarly, a designer will frequently be asked to solve problems the like of which he has never handled before; furthermore, he may know nothing of the client's organisation or markets. Yet the designer must quickly sense the way his client's company 'does things' as well as the finer (or coarser) aspects of the political climate within which his client operates. For his part, the client must endeavour to tie in, wherever possible, with the way the designer goes about his work.

Establishing an effective working relationship based on trust involves much communication. The fact that designer and client may speak different languages could present an obstacle in the quest for common ground. Typically, designers are more articulate visually; clients are more articulate verbally. The terms are broadly similar, the assumptions underlying their use may nevertheless be quite different.

During preliminary discussions, the competitive element is often apparent. Neither party will wish to divulge too much information or offer too much advice; the knowledge and reactions of the other are being tested. Neither party will be keen to demonstrate ignorance or incompetence. Images are being presented. Yet the designer will find it easier to plead ignorance of business matters, for there is generally less expectation that designers will have such knowledge. But because everyone possesses design skills to a greater or lesser extent, it is much more difficult for the client to admit to being ignorant of design matters even if he were to believe it himself. An admission of ignorance could weaken respect and the potential for maintaining control – and clients tend to feel strongly that *they* should remain in control at all times. The client will also seek to negotiate the designer down: budgets will be restricted and deadlines tightened, perhaps because all first suggestions are automatically taken to contain too much slack. After all, no tough and confident negotiator accepts first demands.

Sometimes this kind of competition serves a useful purpose: designer and client get to know each other better, they clarify the problem and agree a project brief. More often it is a superficial charade: an irresponsible waste of time and money which both creates resentment and reduces the chances of reaching a successful conclusion to the project. *Active* cooperation between client and designer normally begins after the designer is appointed and the project is formally sanctioned, though in some instances the parties will keep up a kind of competition throughout the course of the project.

Clearly the selection of designers is a major problem. A survey of management perceptions of the difficulties encountered when handling design projects revealed that, overall, respondents with experience of such projects placed designer selection seventh in a list of twenty-eight suggested difficulties. Those without such experience placed it second (see Appendix A). Designers agree that selecting the right designer for the problem at hand can be particularly difficult. In a complementary survey of designers' perceptions, there was a clear indication that major project management difficulties arise because clients often appoint the wrong designer to projects.

The timing of appointments can also be tricky. Designers complain frequently that they are appointed too late to undertake their work rigorously. Sometimes the client has done the wrong kind of preparation, making premature design decisions and so limiting the designer's scope. Sometimes the client has failed to make any preparation: the problem has not been defined and deadlines are too tight to allow the designer to take on this extra analytical work. Many clients suggest that it is the inherent pace of business and its pervading uncertainties which dictate tight programmes for design projects: designers must accept 'practical' constraints which often lead to compromises, and will have to learn to live with the fact that there is never sufficient time to produce the 'right' solution.

How does the client find out about competent designers for his projects? On what criteria should he select designers? And, what is a typical selection procedure?

Sources of information on designers

Word-of-mouth recommendation is by far the most common means by which clients hear of designers — rather more frequently from friends (and friends of friends) than from business contacts. Suppliers, such as printers, can sometimes be particularly useful sources of recommendation. Comment in newspapers, magazines, radio and television is probably the second most powerful means of exposure. The Yellow Pages are consulted by a relatively small proportion of clients. Surprisingly, the convenient selection services offered separately by the Design Council and the Society of Industrial Artists and Designers (SIAD), though well publicised, would appear to be grossly under-used.

Ideally, the objective of the search for designers ought to be, first, the matching of the particular skills and experience of a designer with the requirements of a diagnosed problem; and, second, the matching of the designer's approach to design problem-solving with the way the client wishes to handle the project. The diagnosed problem and the way the client wishes to handle the project are principal indicators of the nature and scope of the design project. A recapitulation of some of the factors which affect the nature of design projects would be useful here.

The varying nature of design projects

With every design project, client and designer have specific contributions to make and responsibilities to bear. It is the client's responsibility to have his problem diagnosed correctly and to ensure that an appropriate brief is formulated. The designer will endeavour to influence the contents of the brief. If his analysis of the problem and the particular client circumstances suggest that the brief is inaccurate, he should attempt to negotiate amendments. The programme of work forms an important element in the structure of a project: this sets out what has to be done and when — whether by the designer, by the client, or by both working together. Implicit in any working relationship between designer and client is the understanding that the client will provide the designer with relevant information, will ensure that his staff cooperate with the designer throughout the project, and that decisions are made when necessary.

Ideally, clients should present their designers with defined problems from which the designers have to conceive, interpret, and formulate solutions: these represent fully-fledged design projects through which the designers make use of the full range of their skills. There are occasions, however, when the client has a clear idea of the design solution he seeks. Form the designer's point of view, such projects may involve little more than good draughtsmanship. With so little demand for creativity, an 'independent mind' could well be at a disadvantage when dealing with such clients. On other occasions, the client may have no understanding of either his problem or how he is to tackle the project. The designer who gets involved in this type of design project often finds that he has a business, as well as a design, consultancy assignment on his hands. These two types of project represent the extremes of the spectrum — ranging from the oppressively overspecified to the ludicrously open-ended.

Not just a sincere, creative designer

Against this background of differing types of design project, what does the designer offer his clients? He offers *expertise* (in the form of creative, technical, administrative, business and interpersonal skills), *experience* (specifically in design problem-solving), and certain *facilities* (studio/workshop space, technicians and so on).

The nature of each design project and the way the client plans to handle it provide indicators of the necessary mix of skills. For example, a project involving the design of a new range of stationery is unlikely to demand the technical and administrative skills demanded by a graphic/signage scheme for a shopping centre. The restyling of an existing product will probably require fewer creative and technical skills than the design of a similar new product. Extra skills are needed if the designer is expected to put forward design concepts *and* implement the solution. Indeed, different *stages* of a project also call for a different mix: Table 4.1 shows how different skills are brought to bear as the project progresses.

Table 4.1 The varying mix of skills a designer brings to bear on a design problem as the project progresses

Stage	*Skills*
Pre-project	
Analysing client problem and particular circumstances	Creative/business/interpersonal
Finalising brief and work programme	Administrative/business/ interpersonal
Project	
Conceptualising solution	Creative/business
Interpreting solution concept	Creative/technical
Formulation of solution	Technical/administrative
Implementation of solution	Administrative/interpersonal
Post-project	
Evaluation of impact of solution	Business/administrative/ interpersonal
Development of solution	Creative/business/interpersonal

Such skills represent the designer's 'in-house' expertise. But does the designer know how to augment those skills with the services of other specialists when projects so demand? As no designer or design team can hope to cover a comprehensive range of skills or styles, experience of working closely with outside specialists and suppliers (such as researchers, technical experts and particular illustrators) is essential. Normally the wider the designer's experience with other specialists, the wider the range of projects he can undertake effectively.

The experience designers bring to design projects can be broken down into three broad categories: experience of *different problems*, experience of *different client organisations*, and experience of *different markets/industries*. In dealing with different problems and industries, the designer should gain detailed knowledge of particular materials, processes and techniques. In dealing with different client organisations, the designer should develop a feel for the characteristics of certain types of organisations and the markets in which they operate.

Designer types

How do different designers react to different types of design project? Reactions are influenced as much by the designer's approach to problem-solving as by his expertise and experience. Thus, to answer that question it is necessary to describe a spectrum of designer types.

At one extreme, there is the *'strict professional'*: an individual who limits himself to design skills and equivalent design responsibilities. He expects to be thoroughly briefed and supplied with all relevant material *before* he starts designing. If entrusted with the solution of a design problem, he expects to be fully involved from conception through to formulation: pre-digested problems are definitely frowned upon, as are clients who become too involved in the design process. The strict professional's prime concern is the formulation of the 'right' design solution — often developed along purist design lines with minimal concessions to marketing or business principles. His questioning will not dig deeply into the business problem to which the design solution is directed. Involvement with the client company will tend to be at arm's length, in the manner of traditional 'client and trusted adviser' relationships. The strict professional would never take client decisions into his own hands. In sum, the strict professional is by nature reactive rather than a self-starter. He believes absolutely that 'good design' sells itself and so does not market his services.

At the other extreme, there is the *'entrepreneur (design consultant)'*: a professional design consultant who has matured into a business consultant/general manager. Though a design specialist by training, he is equally at home with general management problems. He markets the 'practical mind with a professional approach to business problem-solving'. Few areas are out of bounds to the entrepreneur (design consultant). If necesary he will research a market, design a product to fill a gap, arrange pre-launch tests, formulate the marketing strategy, organise the launch, tie up the outlets, and perform any other activity to make a success of the project. His prime concern is to produce designs which 'work' in marketing as well as in visual terms; if pressed, he is more likely to agree to compromise design solutions – partly because, being conversant with a wide range of business matters, he tends to appreciate the marketing trade-offs sought. The entrepreneur (design consultant) does not freely seek clients who are rigidly conventional in their thinking – particularly where working relationships are concerned. He prefers to go into 'a kind of partnership' with his client, with a joint commitment to solving the business problem which gave rise to the project. This often results in the consultant becoming more involved in the client company, seeking a wider range of detailed information. In the absence of hard data, he will work out informed guesstimates rather than hold up the project while accurate information is still being gathered. If necessary, he will take decisions on behalf of the client. Sometimes, the entrepreneur (design consultant) will actually generate design projects by convincing the client that, on the basis of his analysis, other aspects of his business require attention.

It is obvious that all professional designers would avoid over-specified problems if they could. Such design work might be better passed to technicians, draughtsmen, artworkers, equipment suppliers, or manufacturers. Open-ended problems, the other extreme, are characteristically inappropriate to the strict professional, but are ideal for the entrepreneur (design consultant). Though few individuals display exactly the characteristics of either of these extreme designer types, they give us the basis for a system by which designer, client and problem can be matched.

The matching processes

Earlier it was suggested that when a client seeks a designer for a project, the objective of the search ought to be first, the matching of the particular skills and experience of the designer with the requirements of a diagnosed problem; second, matching the designer's approach to design problem-solving with the way the client wishes to handle the project. Closer analysis reveals that, once appropriate designers have been located, most selection problems are related to four kinds of match-making. These are:

Matching designer to problem: How do the designer's skills and facilities relate to the requirements of the problem at hand? Has he experience of solving just such a problem? How does the designer propose to tackle the problem?

Matching designer to client's particular circumstances: Has the designer ever tackled such a problem against a similar client background? Has he encountered this kind of background while tackling a different type of problem?

Matching client to problem: Has the client diagnosed and understood his problem correctly? Has he decided to handle the project in the most appropriate manner?

Matching designer to client: Do the client and designer respect each other? Do their views on how the project should be handled coincide? Will they work together efficiently?

It may seem strange to include client-problem matching on the list. But inaccurate diagnoses of problems are frequently revealed at the selection stage, and many thorough designers are struck off short-lists because the client reacts adversely to their detailed questioning of the brief. No doubt some potential clients become over-sensitive when the brief is not automatically accepted, or when the designer asks questions about factors not seen to have a direct bearing on the problem. These clients commonly consider the designer to be lacking interpersonal sensitivity and therefore, by implication, skills: they cannot conceive getting on with the designer and, unfortunately, a client-problem mismatch is rationalised (or interpreted wrongly) as a client-designer mismatch.

Another pitfall in the matching process is the rigid association of a problem with the client's particular circumstances. Of course problems do not occur in isolation. Yet the fact that a particular problem occurs in a particular context rarely produces special conditions: the familiar statement from a client that his business is like no other, hence his problems are unique, is inaccurate in nearly every case. The client who seeks only designers who have tackled very similar (or identical) problems for competitors restricts his choice unnecessarily – though he may see this as reducing the chances of failure as well as the time spent on the project, and hence costs.

Normally the client should concentrate on a choice between the designer who has experience of a similar problem and one who has experience of a similar set of client circumstances. It would be uncommonly fortunate to find a designer who qualifies on both counts. If the client decides that experience of similar problems is more important, his area of search may be sharpened to fit the details of the project at hand.

Consider the following case. A toy manufacturer plans to introduce a new product. He has no specific idea as to what this product should be, but his plant has spare plastics-moulding capacity, and he knows there is a gap in the market for battery-operated toys. What is the essence of his problem? What type of designer should he seek?

Stated simply, the client needs a toy designer. However, he may decide that he will be better off hiring a designer, perhaps fresh to toys, who has been notable in introducing successful products to the market. Alternatively, the client may place a premium on knowledge of his particular moulding process because he has experienced major problems when translating inappropriately-conceived designs into production reality. Again, if the client is disturbed by the difficulties which might be encountered in producing an efficient battery-operated mechanism, he may prefer to work with a designer who has designed many battery-powered products.

With the first option, the client interprets his problem as requiring, primarily, product/market-type skills: in this case, he seeks a toy designer or more specifically, say, a toy truck designer (if that is the area he is in). With the second option, the client places greater emphasis on marketing and new-product development skills. With options three and four, the client pumps for technical expertise — but of very different kinds.

With other problems, knowledge and experience of the client's particular circumstances will take precedence over other factors in the matching process. For example, knowledge of a particular market, type of organisation or organisational condition may be an essential prerequisite to solving the problem.

Of course, it should not be ignored that there must be a first time when a designer tackles a particular type of problem, and many designers produce immaculate solutions in these circumstances. Clients do not always examine their designers' previous experience — perhaps because many 'do not understand' design (but know what they like, of course), or because they feel 'approach' is more important than skills and experience. Indeed, clients sometimes assume that professional designers work to acceptable visual standards, and what they seek to determine is the designer's intellectual grasp of the job. This may be a reasonable strategy if the client has a clear grasp of the problem himself. But what if he is confused, as so many clients are? In this instance, he may seek a designer who can carry the project with the minimum of support and supervision — the same designer type who would be appropriate for the client who 'has no time to deal with the project'.

Designer selections based chiefly on 'approach' can create problems. The same is true of selections made with too little regard for 'approach', but the former is especially dangerous if the selection criteria are, in effect, reduced to one question only: does the client like the designer? Is the client taking the soft option by picking the uncontroversial nice guy rather than the challenging loud-mouth who is prepared to comment bluntly on, say, the lousy management in the company? To make the former choice may demonstrate basic psychology in practice, but it can also be irresponsible management which results in 'nothing' designs.

The selection process

When should a designer be introduced into a design project? There are several answers to this question, but only one which is sensible: the designer should be brought in *after* the client has thought through his problem, has defined his problem in some detail *in writing*, and has come to some decision as to how he will handle the project. How else can he determine the characteristics of the problem and project, and the skills needed to deal with them?

The following procedure might be considered as a basis for the selection of designers:

☐The client analyses and defines his problem or need. He prepares a *preliminary* brief and decides how he will handle the project. (For instance, does he favour a tight brief and close supervision of work?) He is now able to draw up a specification of the type of designer he requires, based on what he interprets as the essence of his problem, and whether his particular circumstances take precedence over this, and so on.

☐The client identifies suitable designers. If the client has little or no experience of working alongside professional designers, and knows of none, he might sensibly submit his requirements to both the Design Council and the SIAD designer selection services. Respected business contacts and friends should be approached for suggestions. Suggestions from different sources should be cross-checked wherever possible.

☐Telephone contact is made with each possible designer/group. Primarily, the client seeks basic details of the practice and its work. He also mentions the problem at hand in order to gauge reaction; a discussion may follow.

☐The client draws up a short-list of three to five names. All these designers/groups are visited. Both sides introduce themselves and comment on their background, work, and products/services. The client explains the project problem in some detail; the written preliminary specification of the problem should be to hand, supported by relevant literature on the client company and/or its outputs.

The client should note carefully how each designer reacts to and discusses the problem — note in particular any positive indication of experience and specific knowledge which are of direct significance to the problem.

Furthermore, the client should try to gauge how much he will *learn* by working alongside each designer — not just about the particular design problem and how it is solved, but also about the management of design projects in general. One of the benefits of working with professional designers ought to be the chance to learn from the experience.

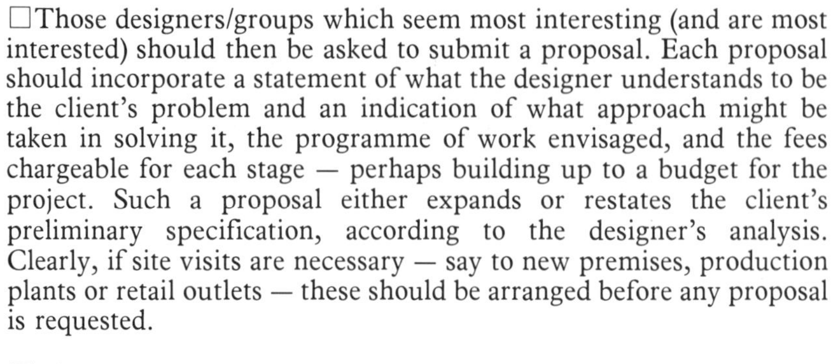

□Those designers/groups which seem most interesting (and are most interested) should then be asked to submit a proposal. Each proposal should incorporate a statement of what the designer understands to be the client's problem and an indication of what approach might be taken in solving it, the programme of work envisaged, and the fees chargeable for each stage — perhaps building up to a budget for the project. Such a proposal either expands or restates the client's preliminary specification, according to the designer's analysis. Clearly, if site visits are necessary — say to new premises, production plants or retail outlets — these should be arranged before any proposal is requested.

□The most interesting proposals should earn the authors an invitation to the client company — to meet other members of staff, to discuss further details of the problem, agree the brief, and finalise administrative matters.

□The letter of appointment should be based on the client's preliminary specification, the proposal of the selected designer, together with any changes and other details agreed at subsequent meetings. These documents should establish the contract between the parties.

This procedure need take no more than three to four weeks, especially if undertaken rigorously and with concentrated effort. The effort put into the selection of designers and the establishment of an effective working relationship between client and designer represents one of the best investments that can be made in design projects.

5 What kind of design project team?

Who makes up a design project team? This question is seldom asked, either by designers or their clients. Sometimes potential clients ask for details of designers who will be assigned to the project if the decision is made to proceed. This is a rare occurrence, perhaps because clients presume that the people they talk to in design groups will be those who undertake the work, just as many clients assume that the range of work shown them was undertaken by an unchanging team of designers. A further common misconception is that the design group appointed alone constitutes the design team, irrespective of the complexity of projects.

Naturally, design groups are subject to similar operational pressures as any other business. They can be particularly vulnerable for most are small compared with other businesses: sizes range from partnerships of two to companies of up to one hundred employees. Units of eight to twelve are considered 'medium size'. To be profitable, most design groups are 'run lean' and operate close to full capacity. During slack periods, designers are laid off. When work is plentiful, freelance designers might be used to cope with what frequently turns out to be a temporary overload. In good and bad times, designers seek change or are lured from one group to another. Thus turnover of staff within design groups can be high, and it need take no more than the addition or subtraction of one or two members to bring about a radical change in competence, approach and capacity. Even with stable groups, the combination of individuals assigned to projects varies. Thus the choice of a particular design group does not necessarily secure a given design team or a given style of solution.

If design teams have wider significance, who else should be considered part of such a team? Are all design teams made up of the same kinds of members, or are there variations with type of project? To answer these questions, it is necessary to discuss some of the fundamental roles encountered in design projects.

The fundamental roles encountered in design projects

The most obvious role is that of the *designer*. Clearly, he is a design *doer:* it is up to him to conceive and produce design solutions. Yet, unlike the craftsman, a designer does not always 'produce' his designs. Few graphic designers are also printers; few product designers have production facilities in-house; few exhibition and interior designers implement their design schemes on site themselves. Thus a second role in design projects is that of *design supplier*. The term does not apply only to those who produce the final designs, but applies equally to those who supply components of the final design. Thus, several design suppliers can contribute to a single design project. For example, typesetters, artworkers, and printers could be design suppliers to graphic design projects, while draughtsmen, model makers and toolmakers are possible design suppliers to product design projects. The term also applies to those who supply data and information, such as market researchers and technical experts.

The *design manager's* role centres on the organisation and day-to-day administration of design projects. This narrow 'mechanical' interpretation of the management role is deliberate. Normally, a manager carries sufficient authority to be responsible for the outcome of projects he manages. With design projects, the person given the responsibility for day-to-day administration is frequently not given the necessary authority to undertake his work effectively, in that he can neither set the budget nor have the final say. In certain circumstances, he is essentially a middle-man and suffers from the inflexibilities and conflicting pressures common to those who fill intercalary roles.

Wider management responsibilities, such as the formulation of design policy and the establishment and maintenance of design standards, should, therefore, fall on the shoulders of *design responsibles* — the controllers of investments and the final decision-makers in design projects. Whereas the design manager's principal role is to organise, the design responsible's principal role is to provide leadership.

The differentiation of these fundamental roles enables us to analyse in greater detail the way in which design projects are organised and managed — particularly with regard to the composition of project teams, and the apportionment of work and responsibility.

The hierarchy of roles

There are several points to be made about these roles. First, all the roles exist in every design project, whether they are acknowledged or not. The contribution each makes varies with the type and importance of the project. Thus an important, fully-fledged design project — such as the design of a brand-new product — will normally draw full measures from each of these roles. With routine design projects and those in which the solution is strictly prescribed — such as the addition of a minor item to an established range of stationery — the roles of design responsible and designer may be less prominent.

Second, there is a natural hierarchy in these roles. Ideally, the design

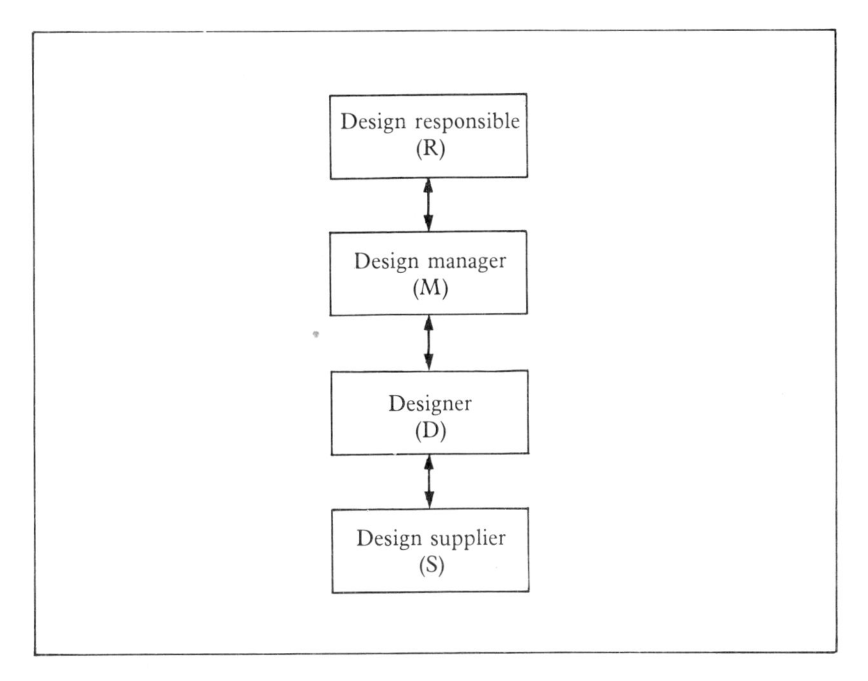

Figure 5.1 *The hierarchy of roles in design projects*

responsibles should have overall responsibility and control of projects. Under the design responsible will be the design manager who administers the designers who, in turn, should be placed above design suppliers.

Third, these roles are not mutually exclusive: two or more roles can be fulfilled by a single participant. Thus design responsibles frequently manage design projects, as do designers; design suppliers can undertake design work and sometimes even manage projects, and so on (Figures 5.2-5.4).

Fourth, the hierarchy of roles suggests a sequence in which the roles should be filled. Normally it will make little sense to appoint design suppliers before designers because the choice of suppliers is influenced by the design solutions produced and the way the designer works. Similarly, the appointment of designers before managers can lead to unnecessary friction.

Finally, the incumbents of roles must be well matched so that they work effectively as a team. The matching process involves considerations of personalities, skills, ways of working, efficiency in communication, and standards.

Developing the project team

The hierarchy of roles gives only the basic configuration of roles.

Most design projects involve more than one design responsible, designer, and supplier. Sometimes projects also involve more than one design manager. Design project teams can be developed in several ways, depending on the problems being tackled and the way the projects are handled.

To start with, several design suppliers are used in the majority of design projects. Thus product and interior designers may use draughtsmen, illustrators, model makers, materials and process specialists, as well as specialist suppliers, in the course of their work; graphic designers may use photographers, typesetters, artworkers, plate makers, and different types of printers and finishers in their work (Figures 5.5-5.7).

The suppliers used in design projects are of two kinds: those who are appointed direct to the project by the designer or client, and those who handle work sub-contracted out by the direct suppliers. Sub-contracting of work may be declared openly; sometimes it is deliberately hidden. The printer may sub-contract out the colour separation work on photographs and the preparation of plates for a prestige brochure. When the brochure is printed and collated, he may send it out again for binding. In this instance, the configuration of the project team is in Figure 5.8.

Suppliers may be commissioned to undertake design work. This happens most frequently when a client is used to dealing direct with suppliers who either offer a subsidiary design service or sub-contract out the design element. Exhibition system manufacturers are an example of suppliers who offer such a design service; printers are another. It also happens in the case of specialist suppliers who probably know more about design problem-solving in their fields than most non-specialist designers. Thus an electronics firm may be asked to design an advanced component for a new product. In these cases, it should be clear that the design solutions offered may be dictated as much by the suppliers' products and facilities as by the client's needs (Figure 5.10).

If the technology offered by a supplier is of primary importance to the work in hand, then there is a case for appointing the supplier before the designer (as the supplier may be perceived to be more important than the designer, it would be more appropriate to choose a designer who will work well with the supplier, as well as being sufficiently knowledgeable in that field). In any case, suppliers at the forefront of their fields probably work regularly with a number of independent specialist designers and so they may put forward names for selection. On occasion, a particular supplier's contribution to a project is considered so vital that the supplier may be appointed project manager as well as undertaking design work. Interestingly, this kind of arrangement is also made when the problem is perceived to be straightforward and relatively unimportant (Figure 5.11).

Thus in the former case, a computer manufacturer may be given the

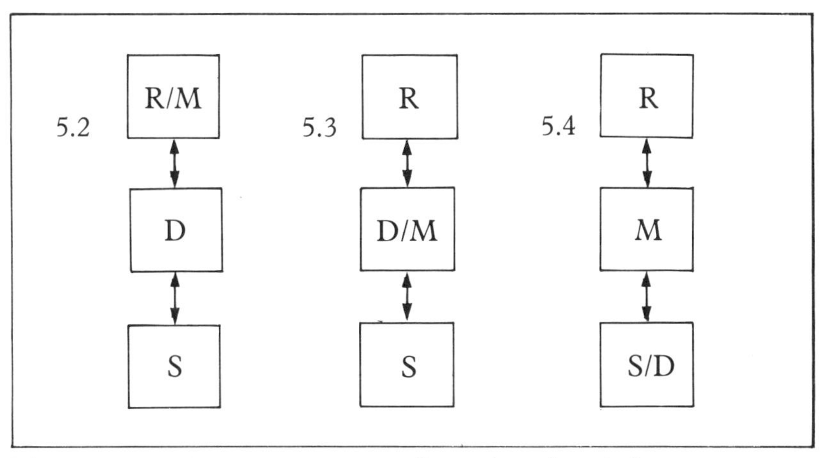

Figure 5.2 *Design project team configuration when design responsible also manages the project*

Figure 5.3 *Design project team configuration when designer also manages the project*

Figure 5.4 *Design project team configuration when supplier also undertakes design work*

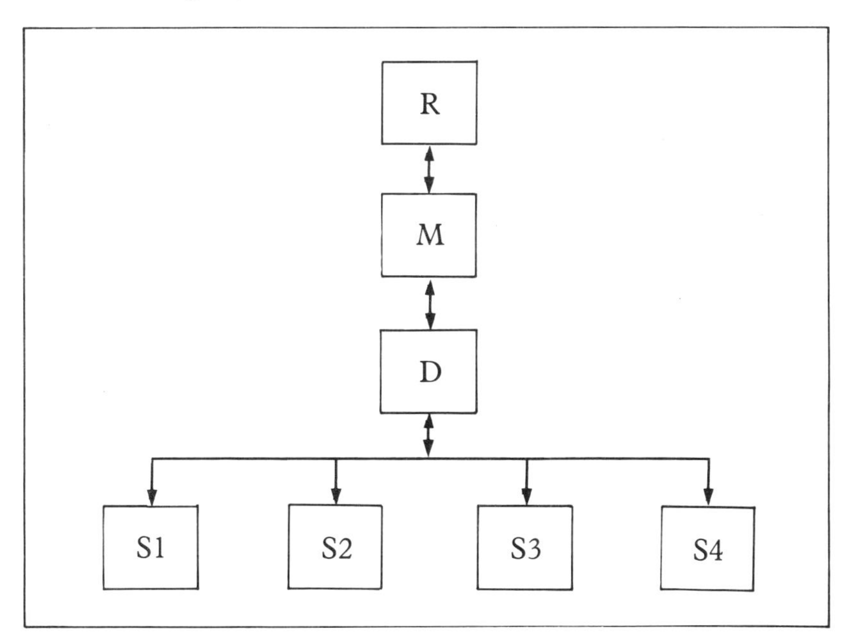

Figure 5.5 *Development of team configuration to take into account the use of several suppliers*

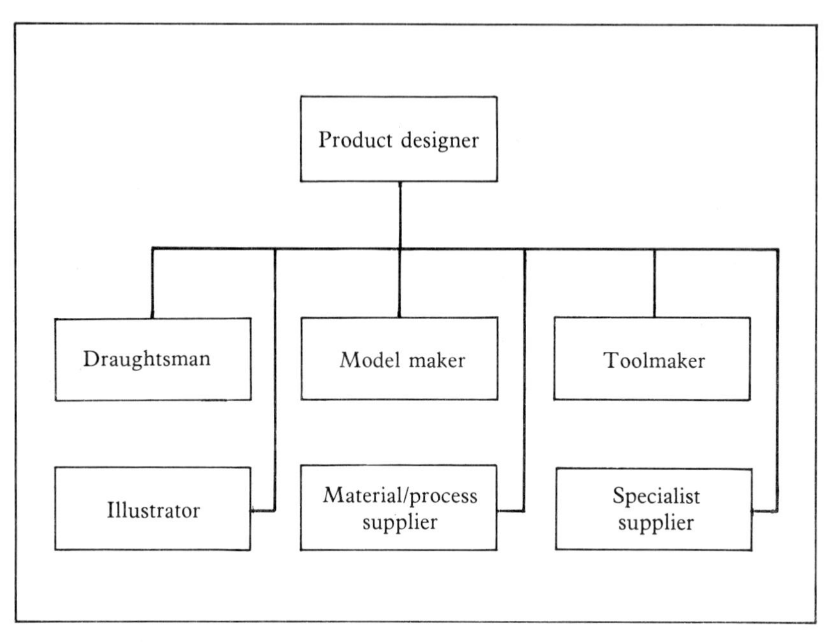

Figure 5.6 *Various suppliers used during the course of product design projects*

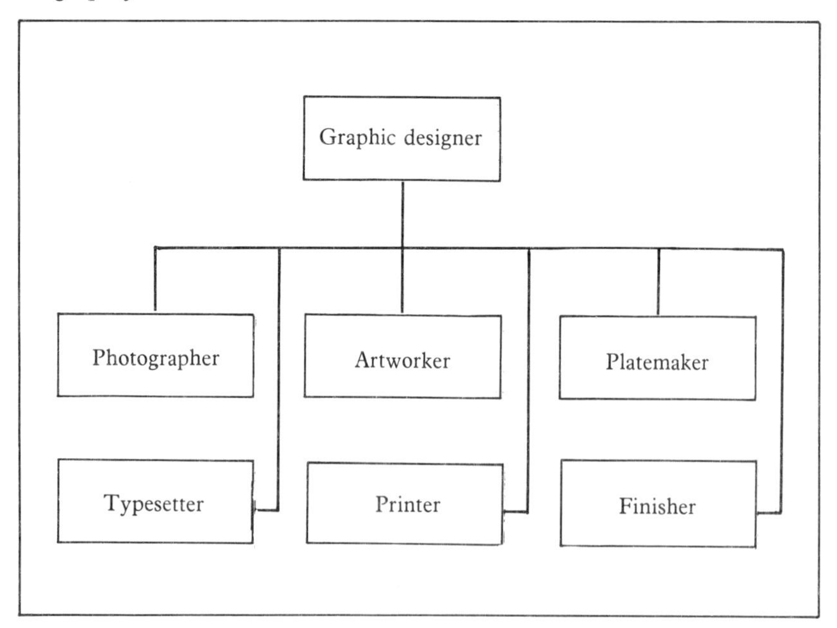

Figure 5.7 *Various suppliers used during the course of graphic design projects*

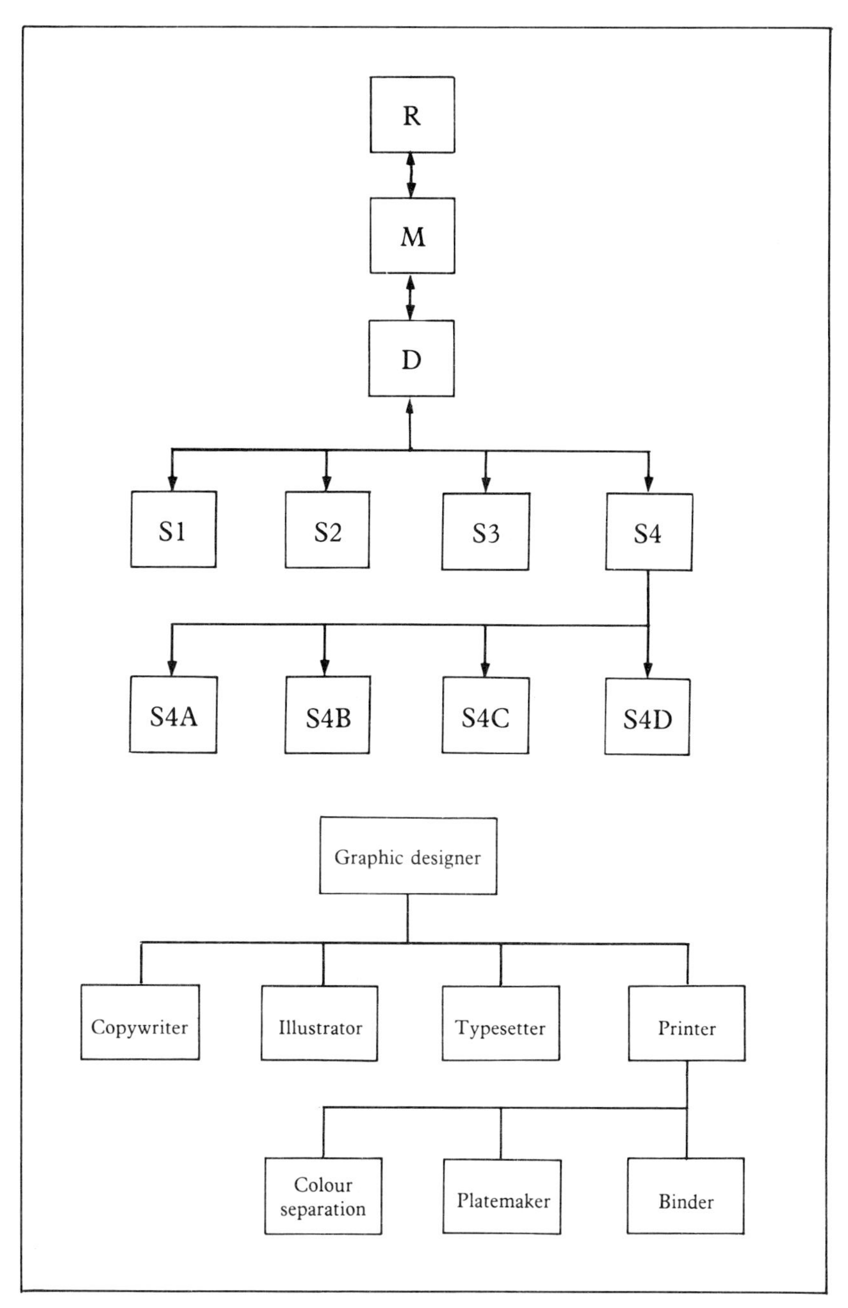

Figures 5.8 and 5.9 *Development of team configuration showing use of suppliers appointed direct to the project and suppliers who undertake sub-contracted work*

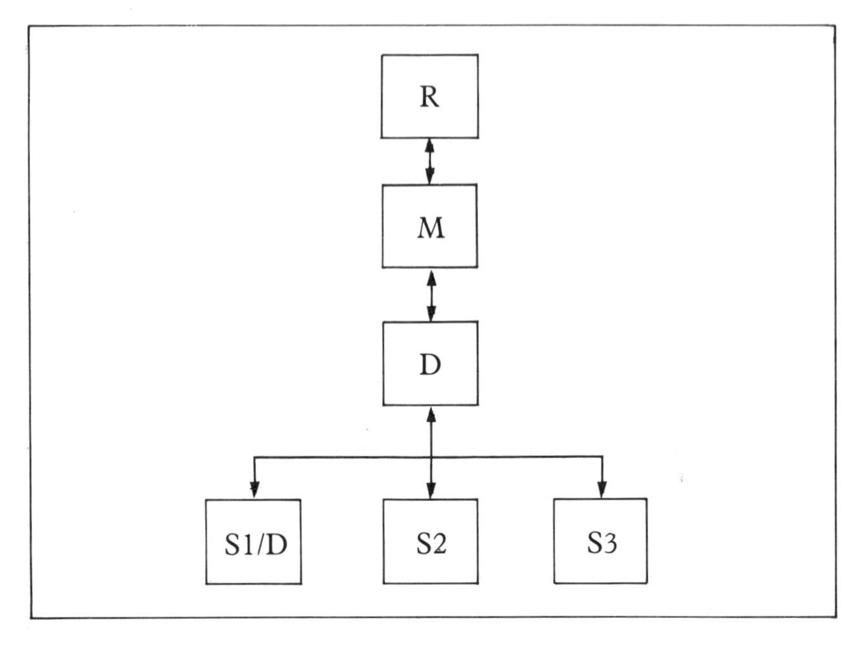

Figure 5.10 *Design project team configuration in which one of the suppliers used undertakes some design work*

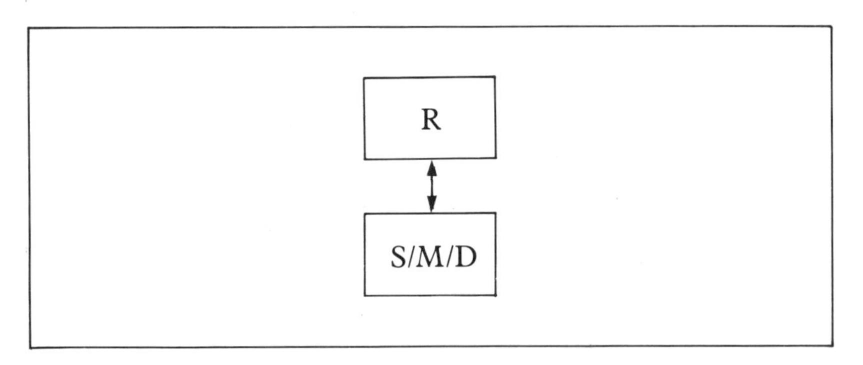

Figure 5.11 *Design project team configuration when the supplier undertakes design work and manages the project*

task of designing and having fitted out the interior environment to house a new installation for a customer. In the latter case, a builder may be asked to carry out minor structural alterations and redecorate parts of an office with minimal supervision.

Several designers may be used on a particular project — either as equals or with one as the superior. In some instances, these designers could be members of one design group or an in-house design department; in others, they may never have worked together before.

Thus in an exhibition design project, the client, acting as design manager, may coordinate the contributions of an exhibition designer and a graphic designer. Alternatively, during the development of a shopping centre, the graphic designers working on the design of the centre's corporate identity and signage schemes will be subordinate to the architects, whether as full members or advisers. In either case, the designers could be either on the staff of the client company or independent practitioners (Figures 5.12-5.14).

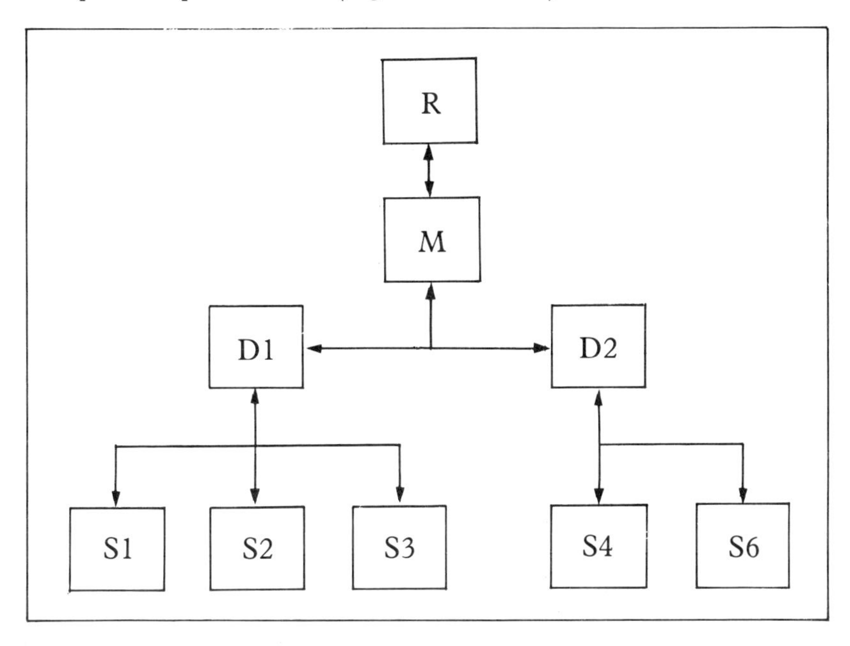

Figure 5.12 *Team configuration when two designers of equal status are used*

The larger the number of designers and suppliers used on a project and the greater the numbers in the hierarchy of roles, the greater will be the administrative burden for each level up the hierarchy. The design of a letterheading will involve relatively little administration; by contrast, in the course of complex projects — such as the design of large suites of offices — designers carry heavy administrative loads associated with their design contribution. Unless the designers concerned are also efficient administrators, an effective solution is unlikely to be created or implemented.

As more complex, longer-term problems are tackled in design projects, the work load will be broken down into smaller, more manageable elements or stages. As project teams increase in size, they too may be broken down into sub-units — the most common being the splitting-off of designers and some suppliers into a 'design team' and

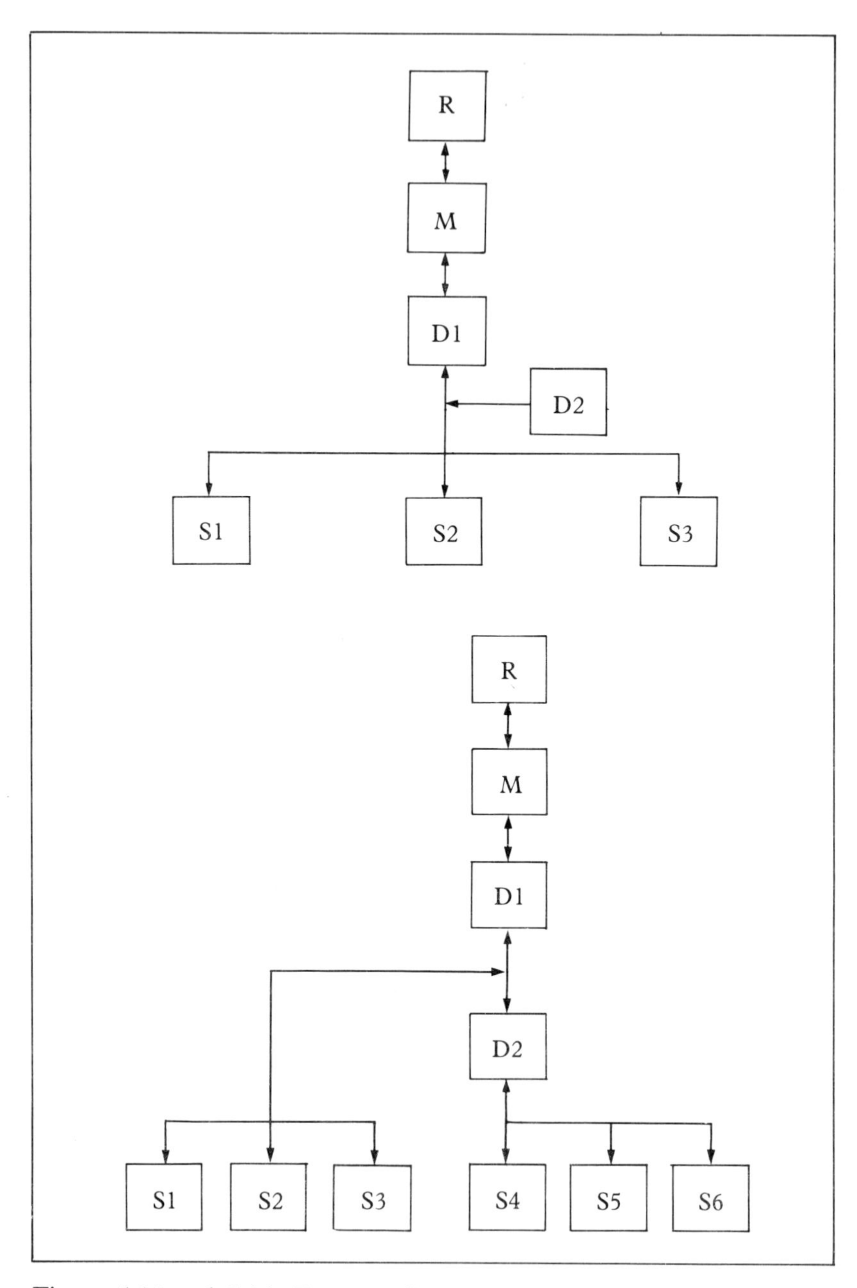

Figure 5.13 and 5.14 *Team configurations when a second designer is brought in either as a consultant to the project designer or as a subordinate*

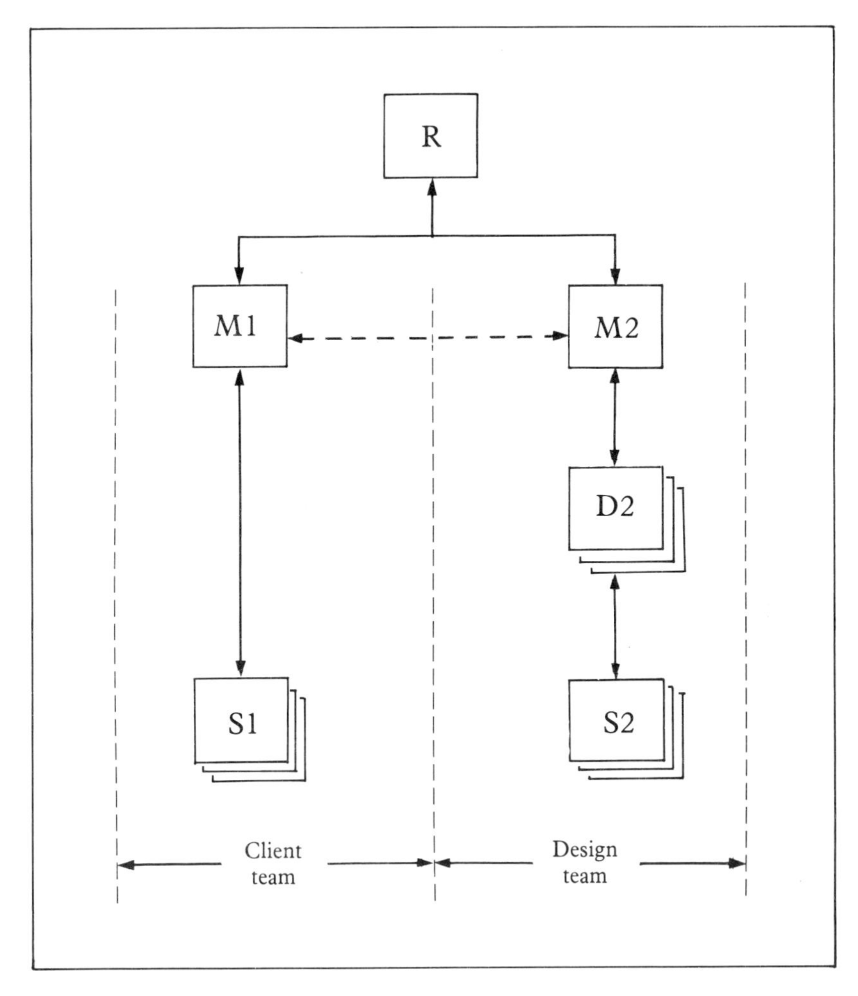

Figure 5.15 *Splitting the design project team into a 'design team' and a 'client team'*

the other participants into a 'client team'. This arrangement, which also occurs with projects with small teams, can bring about a fragmentation of the managerial role, and the danger of a diffusion of responsibility within the project team. A different manager may be formally given the task of administering separate elements of work; alternatively, one manager may take the project from initiation to the formulation of a design solution, while another takes it through the implementation stage.

The proliferation of design managers within a project may not be deliberate. Several design managers may emerge because the job of project manager has not been clearly assigned. If the person put in

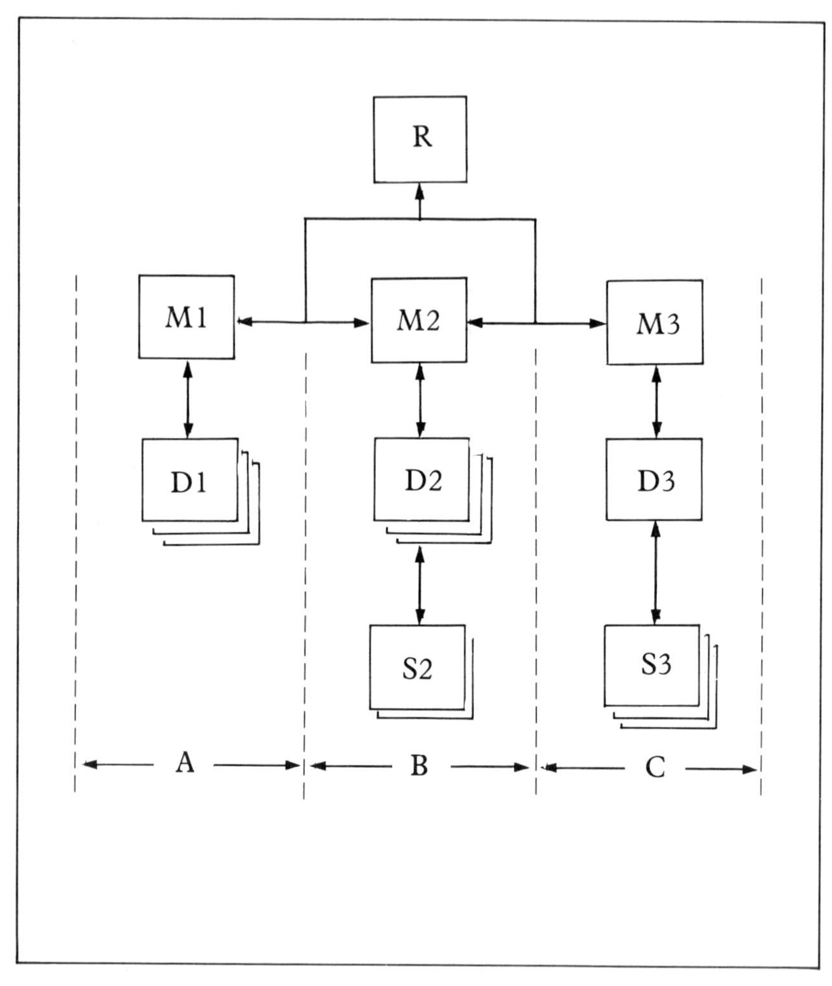

Figure 5.16 *Splitting the management of projects according to aspects of work*

charge of a project is unable to cope or is unwilling to make a full contribution, team members may seek guidance elsewhere. In all these circumstances, leadership and coordination become critical. The design responsible may take it upon himself to coordinate the various activities within a project. Alternatively he may appoint one of the managers to the task: this individual then rises in status to the equivalent of project manager, and a further level is added to the hierarchy of roles. If the design responsible is unaware of the way the project is being handled day-to-day, the task of coordinating activities will fall by default onto the shoulders of another team member – usually the designer, who then has to carry the project. Again, a level is added to the hierarchy of roles as in Figure 5.19.

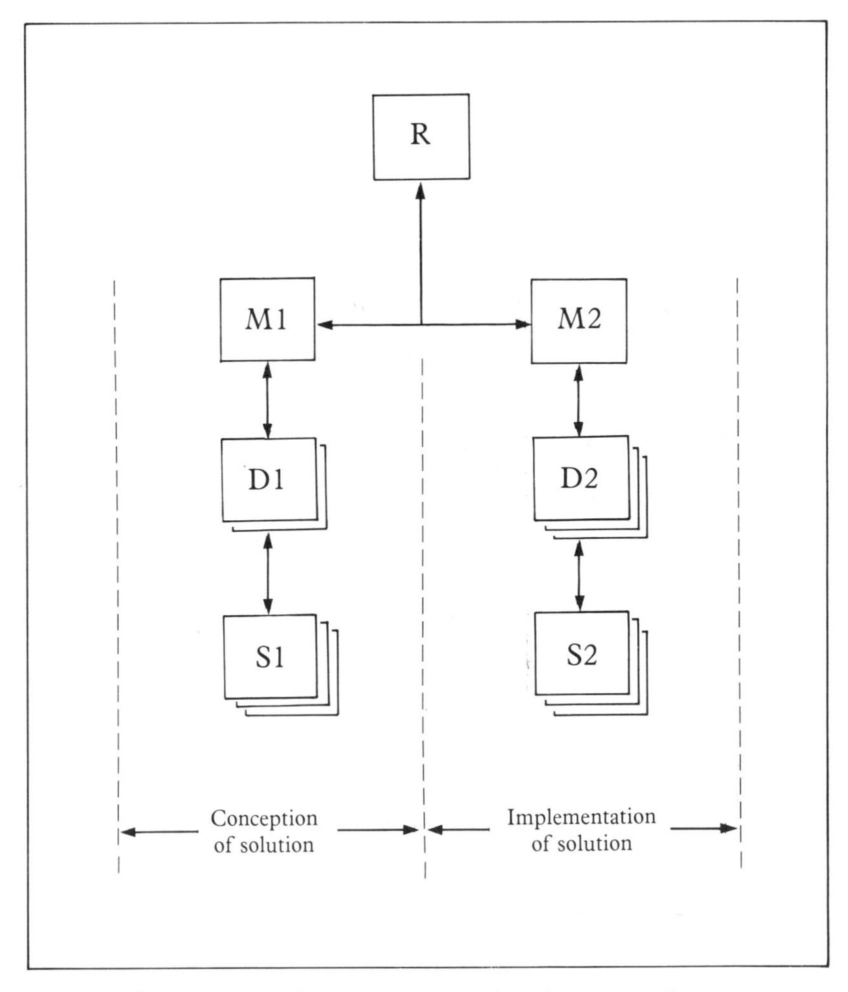

Figure 5.17 *Splitting the management of projects according to stages of work*

Confusion may particularly occur when a design project is 'managed' by a committee. The committee may be made up of individuals relatively low down the management hierarchy and reporting to a middle manager, or it may be a high-powered group consisting of board members and senior managers, and reporting directly to the board. In theory, the committee chairman ought to be the design manager. Through his control of the committee, he should exercise control of the project. Yet this is not always the case. Committees appoint members other than the chairman to deal with project administration, and the ways and influences of committees should never be underestimated. If the chairman is not a design responsible and does not pursue the project actively, then he

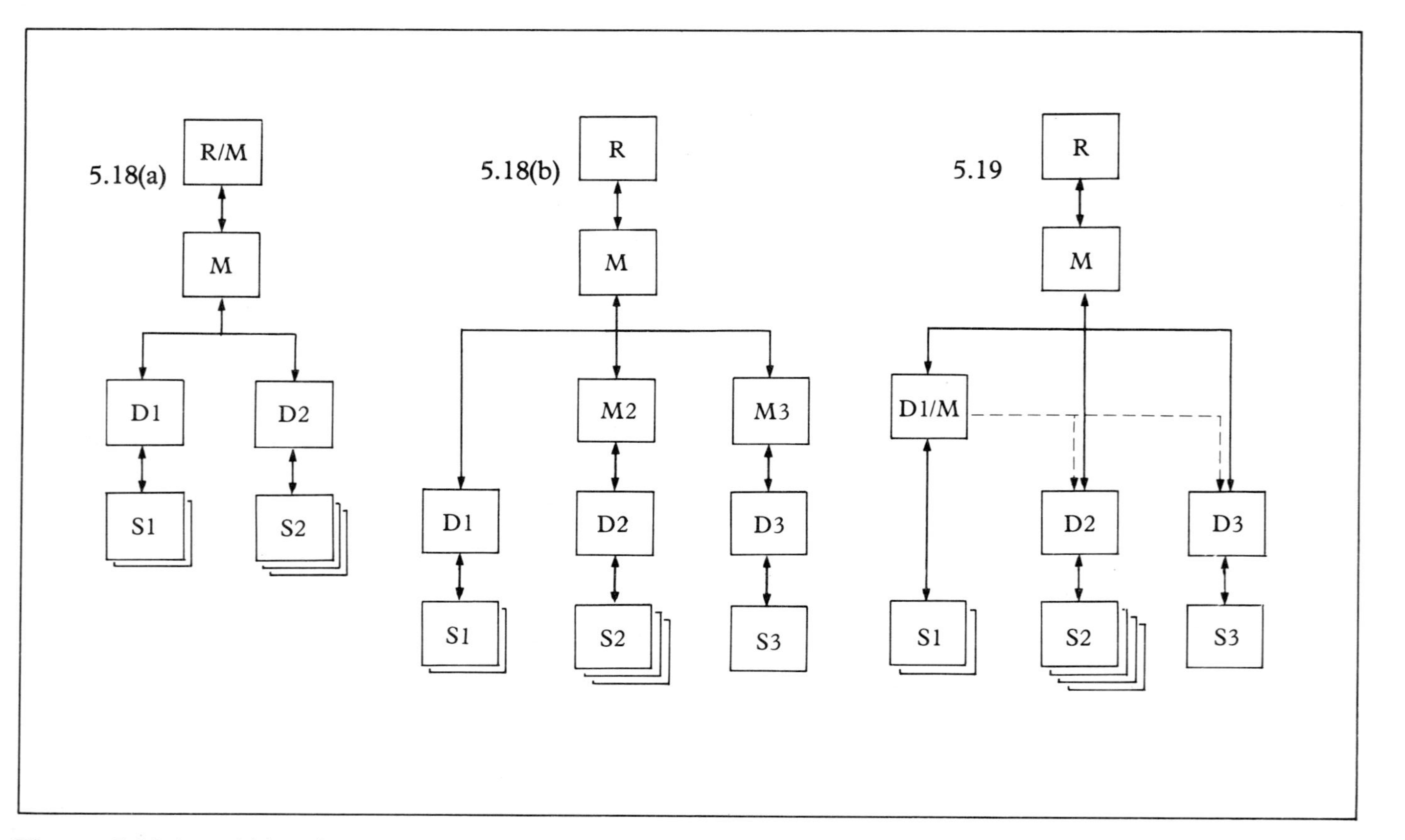

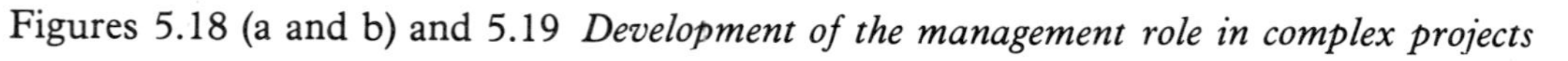
Figures 5.18 (a and b) and 5.19 *Development of the management role in complex projects*

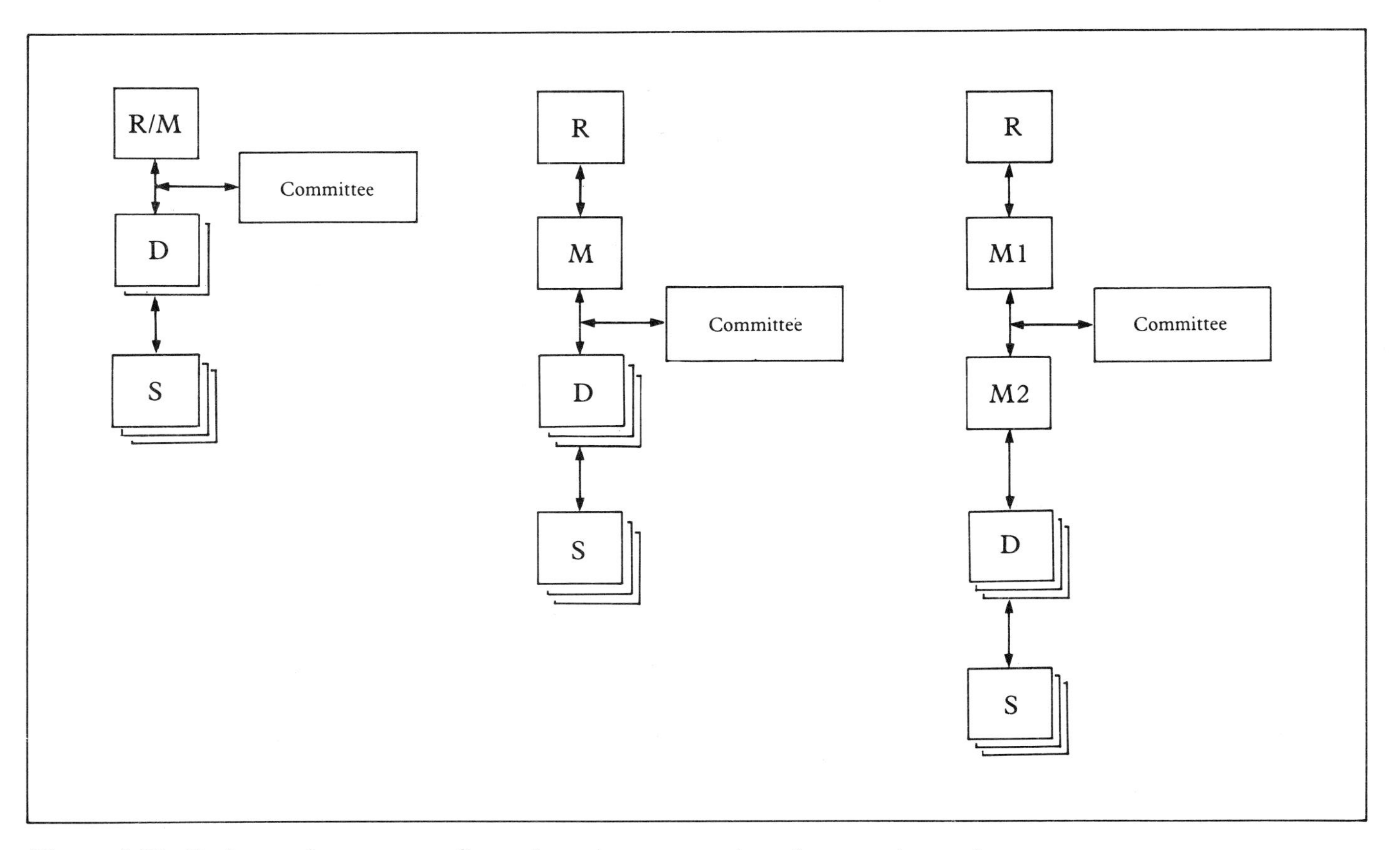

Figure 5.20 *Design project team configuration when a committee 'manages' a project*

effectively abdicates his managerial role and a further level in the project team hierarchy is created.

Mapping out team structures in this way and relating them to the project work set provides a means of determining whether there are sufficient members in the team to undertake the work properly and whether these members possess the necessary skills. The analysis is particularly useful as it also gives some indication as to how the various team members contribute to the progress of projects. To gain a fuller picture, the way project teams operate needs to be analysed.

How project teams operate

The way a team works will have a strong effect on communications within the team and the way decisions are reached.

At one extreme, projects are operated in a sequential, 'mechanistic' mode. This involves the design responsible and design manager agreeing the objectives of the project first. The design manager then briefs the designer who works on the problem, obtaining information and quotes from suppliers, before presenting his solution to the design manager; if the manager approves the proposals, he seeks the go-ahead from the design responsible. The stages of work are taken in strict sequence. Thus projects will start with research, proceed through the structuring of problems, conceptualisation of needs, interpretations of need concepts, formulation of solutions, and end with implementation of solutions generated. At each stage, the decisions made are aimed at limiting the options in succeeding stages. In the ideal project, the progressive 'firming up' that occurs would lead to a single, most appropriate solution — a very rare occurrence indeed! Furthermore, where the project stages are handled by different people or departments, the contribution each makes is considered complete and final at the conclusion of each stage.

This mode of operation is often accompanied by sequential communication up and down the hierarchy; it also tends to involve a rigid chain of command. Responsibility and authority remain firmly at the top. Direct communication takes place only between adjacent levels of the hierarchy: designers are allowed only indirect access to responsibles through managers; suppliers only gain indirect access to managers and responsibles. Communication between team members and others, whether inside or outside the client organisation, tends to be channelled through the design manager. Thus there is limited opportunity for cross-fertilisation of ideas, and it could be argued that this mode of operation does not involve a team at all, but rather sub-groups of individuals.

Such an approach tends to suit routine, short-term projects: those in which the work is highly structured, and where solutions are not expected to be out of the ordinary or are predetermined. In these cases relatively small, single-discipline teams or individuals are involved,

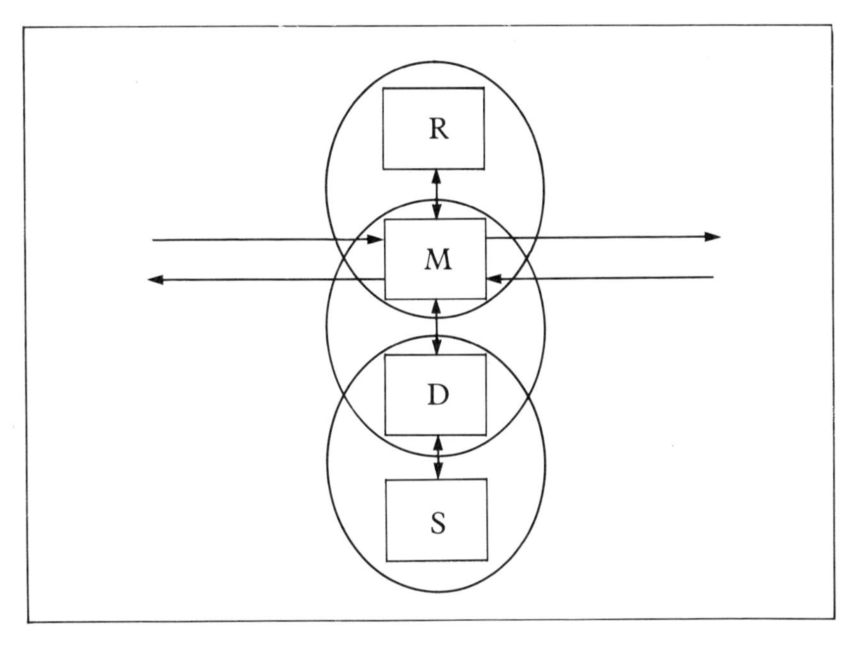

Figure 5.21 *Sequential, 'mechanistic' mode of project team operation*

the work tends to be familiar and instructions straightforward. Wide contacts inside or outside the team are not essential. Designers who are of the *'strict professional'* type tend to favour this mode of operation because it allows them to concentrate on design problem-solving.

At the other extreme, project teams adopt an integrated, 'organic' approach, meeting regularly *as teams* to work through problem areas and points of detail. At such meetings, responsibles, managers, designers, and suppliers have the opportunity to contribute equally to the project *as a whole.* In between times, specific elements of work will be allocated to sub-groupings of the teams.

Communication within the team and with the outside is widely based, the mode of operation relatively fluid. The stages of work are not necessarily taken in sequence: several stages may be worked on concurrently; solutions are evolved through an iterative process between stages. Tentative decisions will be taken and the implications examined across the span of stages. A number of solutions may even be formulated, then tested, before a final decision is made. If different departments handle different stages, their involvement in (and contribution to) the project is rarely confined to those stages, but may straddle several stages or even the total span of the project.

Though the chain of authority is clearly defined and ultimate responsibility for the outcomes of projects remains firmly with the responsibles, responsibility for elements of work may, nevertheless, be

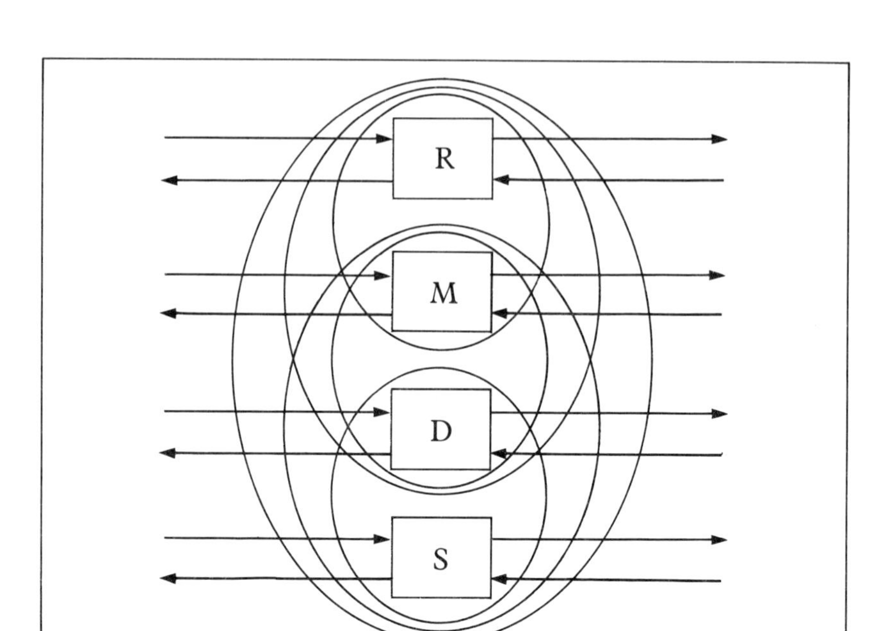

Figure 5.22 *Integrated, 'organic' mode of project team operation*

vested lower down the hierarchy. This mode of operation is best suited to projects which involve complex, open-ended problems which may be unfamiliar or very advanced technically; where cross-fertilisation of ideas is critical and persuasion plays an important part in shaping both problems and solutions; where the outcomes are anticipated to have important, longer-term impacts on client organisations. Such projects tend to use relatively large, multi-disciplinary teams which include designers who are of the *'entrepreneurial (design consultant)'* type.

Who should fill the team roles?

Getting the structure of a design project team right is obviously important: bringing together individuals who respect one another and can work effectively together within a team framework cannot be over-emphasised. Yet there tends to be little thought given to the question: 'Who should fill the various roles in a project team?' The selection of the right designers for different types of design project is a major problem area which worries managers (Appendix A). Nevertheless, the selection process is rarely undertaken rigorously, and many managers admit that had they approached the task more professionally, a larger number of designers would have been interviewed and a different designer would almost certainly have been appointed. The selection of suppliers is not much easier.

Of crucial importance is the selection of members of the client organisation to fill the roles of design manager and design responsible. Agreed, this is not always a matter of choice: in the majority of companies there are rarely more than a handful of managers who can be allocated these tasks and responsibilities. However, where there is a choice, that choice ought to be made with the utmost care because the outcome of design projects is often determined by the calibre of design managers and design responsibles.

Whereas no company can abdicate ultimate design responsibility to an outsider, all companies have the option of using outsiders to manage projects. Indeed, many established design groups, and especially those which handle large projects, have members whose principal task is to administer projects. Entrusting such individuals either formally or informally with the management of design projects can be a tremendous convenience to clients.

Logically, the more important a project and the wider its anticipated impact, the more senior the design responsible and design manager should be within the client company. Yet many projects which are vital to the future profitability of the client company are entrusted to junior executives, and it is not uncommon to find senior managers dabbling with trivial problems.

Are there any guidelines to which one can turn, given that no two companies are identical and that there are different types of design project? Yes, there are.

First, all decisions should be made at the lowest possible level in an organisation — as close as possible to the point where project work is being undertaken and where the outputs of the project will be used. Care must be taken to ensure that persons given the authority to make decisions also have knowledge of the objectives and implications of these decisions, or have access to this information. Thus, familiar or routine problems whose outcomes are predetermined, confined in scope, or are easily and cheaply reversible, can be dealt with quite low down an organisation. For example, an office clerk might supervise the minor amendment of a business form; a product manager might be wholely responsible for the updating of the product, perhaps even for introducing a minor extension to the product range.

Unfamiliar, technically complex problems whose outcomes are anticipated to have wide and long-term effects should be dealt with by senior members of the client company. Apart from the relative importance of these projects, the 'fluid' mode of operation often adopted with such projects requires the skills and authority of senior executives to maintain tight control.

Problems which call for the active involvement of staff from several departments or divisions of an organisation will normally require senior executives in the roles of design responsible and manager though, again, this is not necessarily the case with routine and relatively unimportant projects. When a company moves head office, a board director should take charge of work on the design and fitting out of the new premises, as well as the organisation of the move itself; or when a company seeks new product concepts which might be produced and marketed five or ten years hence, the task of marshalling efforts in this search ought to be the responsibility of the managing director, not of a product manager. Similarly, the conception of a corporate identity is the responsibility of the company board as a whole and not of any single member.

Second, before selecting members of a project team, it is worth analysing the breadth of impact anticipated within the organisation. For the objectives of choosing a project team must encompass the long-term survival of the solutions implemented as well as the efficient progressing of work leading up to a satisfactory outcome. What demands will the project make on various sections of an organisation? Who will the project manager have to deal with during the course of the project? In order to ensure acceptance of the project, could a member of a particular department be made part of the team rather than being consulted informally? Alternatively, should outsiders be appointed to the team even if the company has the required skills?

Clearly the danger of ending up with a large and cumbersome team is always present. Yet experience suggests that many projects are set up with teams too small to undertake the work properly. By the time this shortcoming is discovered, it is frequently impossible to expand the team formally as budgets are inflexible. Consequently, informal assistance is arranged which cannot be subjected to rigorous control, and the project suffers.

The balance between formal and informal team members is a delicate one. So is the balance between those members who are selected principally for their technical skills and knowledge, and those who gain places because it is politically expedient for them to be involved. There are two key issues concerning political team members: is it more important to gain acceptance of the work being done on a project while it is progressing, or can such acceptance be gained when solutions are recommended? Will inclusion of political members facilitate (or even improve) the project work sufficiently to counterbalance any inconvenience caused by their involvement? Naturally these key issues can only be dealt with by examining projects individually.

The more politically sensitive a project, the more senior the design responsible and manager should be. Indeed, if the design responsible is sufficiently senior in rank, his involvement in the project may do away with the need for formal political team members altogether.

The appointment of outside designers can also be a politically charged decision in companies which have in-house design facilities. Normally outside designers would be used if the project were seen to require skills which do not exist within the company or if staff designers are fully committed to other projects. But outside designers are introduced into project teams in order to raise standards and to inject new thinking when fresh styles of solution are required. For some companies, this is the most rewarding way of keeping in touch with design trends and technical developments.

There are circumstances in which it becomes impossible to use staff designers, perhaps because conflicting departmental interests are seemingly irreconcilable. The appointment of a 'neutral' outsider often appears to be the most convenient course of action.

Finally, there are projects considered to be so prestigious by the companies concerned that only independent designers with established reputations are approached. In many such projects the operational and marketing benefits of using famous designers are indeed great. With others, the importance of the project or the potential contribution of the designer is exaggerated and expectations remain unfulfilled, both in terms of management experience and financial benefits.

An essential need: a team approach to design problem-solving

Perceptions of 'Them' and 'Us' are not confined to managers and shop-floor workers: they apply equally in working relationships between managers and the specialists they use. Managers often perceive a loss of control when problems are 'packaged' and 'handed over' to specialists. This is not helped by the fact that many specialists (and especially designers) 'take the problems away' for solution. Of course, in reality problems do not move away from the client organisation unless they were not there in the first place. And the responsibility for solving problems cannot be handed over to anyone outside the organisation. Nevertheless, it *seems* that way to the managers concerned. So, while the specialists work away devising solutions, managers may feel they have little to contribute to the work: tension builds up as they wait for the specialists' pronouncements. Alternatively, managers sometimes look upon specialists as necessary but unwelcome intruders: they will prescribe and control the specialists' work closely, allowing minimal discretion. Again, tension builds up between the parties.

In both cases the tension is destructive and acts against those involved in such projects working together as an efficient unit.

This analysis of roles and team structures has hopefully clarified *some* of the factors that have to be considered when building up design project teams. But more important still, we hope it has emphasised the essential need for a team approach to design problem-solving — an approach which knits together the interests of managers and designers. After all, though outstanding design solutions are often the fruits of individual vision, their acceptability and effective implementation often emerge from successful team effort.

6 So you think you need a corporate design manager?

Since 'design management' hit the headlines a few years ago, the debate on why design standards are so low in this country has intensified — which is not to say either that it rages furiously or that it is taken seriously by industrialists. Periodic correspondence in national newspapers tends to reiterate the polarised opinions of businessmen and designers. Unfortunately there appears to be no real interest in establishing common ground between those who design and those who pay for design.

That industry is generally not professional in its approach to design and fails to harness the design talent available can be seen to some extent in the quality of goods produced and how these goods are presented to the market; it is also evident in the quality of work environments up and down the country. What is not at all evident is *why* design standards are so low.

For years it was considered that the quality of our designers was the principal determinant of the quality of design: because designers were insufficiently trained to meet the demands and rigours of industry, standards suffered. Inevitably there is some truth in this view where college graduates and junior practitioners are concerned. But no profession should reasonably be judged by the calibre of its raw graduates and junior practitioners. Overall, the design profession probably contains a proportion of incompetents similar to any other profession. However, this nation has produced more than its fair share of outstanding designers over the centuries and, today, the wealth of talent available is the envy of many of our trading partners. The past two decades have seen an unprecedented demand from abroad for the services of British designers, and many British companies now have the added challenge of competing against foreign companies which manufacture products designed by British designers.

Support for the argument that the causes of the low standards should be sought outside the design profession came from the findings of a recent survey on management perceptions of the difficulties encountered when handling design projects (Appendix A). These findings revealed that managers — or, more broadly, clients — are uncomfortable about their roles and skills in managing design projects, not least because they lack understanding of the nature of such projects. Specifically, there was strong agreement among respondents in Britain and Canada that difficulties were created because management does not prepare itself rigorously enough when undertaking design projects and because senior managers do not appreciate what such projects entail. Surprisingly, there was no agreement that a lack of commercial orientation in designers leads to difficulties. Given that managers represent the major market for designers' skills and that they are in virtual control of which designs get to the market place, these findings are particularly disturbing.

Should designers be appointed to company boards?

If low standards in Britain are principally a 'management problem', what should be done about it? Some suggest that the problem is structural in nature: that there is little hope of a general improvement in design standards unless design is accorded a status equal to marketing and production, and design activities are controlled by company boards.

Experience does not confirm that such changes are a necessary foundation to improved standards in all types of organisation. For often it is in the day-to-day management of design projects — typically entrusted to those much lower down the management hierarchy — that design standards are established and maintained. Furthermore, as the survey findings indicate, in many companies it is senior management which fails to grasp the potential contribution of design.

So is the answer to have designers on company boards? This vision has many attractions. Design would be recognised formally as a separate and critical function, designers would gain positions at the highest levels of industry, and design in industry would be controlled by trained specialists. Unfortunately, it is likely to remain a vision in the foreseeable future for, though we may have many outstanding designers, this does not necessarily mean that they can operate on equal terms with other directors on company boards. The number of designers who could currently serve effectively as board members is probably embarrassingly small. And of those capable few, a proportion will prefer to retain their independence of 'client' companies. There are professional reasons for this.

As yet, 'design management' has no status among designers. Practising designers who are also heavily involved with project administration are not seen to be 'managing' but simply practising their profession. In these cases, the principal management tasks are expressed as the programming of work, the provision of guidance and inspiration to designers, and the protection of design teams from excessive client demands and disruptive interference. When trained designers cease to practise and concentrate instead on administration, they are likely to be considered failed designers and written off as nonproductive overheads. There is the added danger that with greater exposure to management pressures and values, the designer-turned-manager may gradually become less sensitive to other designers. If he then tries to relate the design team too closely to the rest of the client organisation, he also runs the risk of being labelled the 'client's boy' and thus lose status within the design team. Progression into design management is not a natural consequence of successful design practice. Designers still perceive 'design management' to be what 'outsiders' do when they become involved in design projects: in essence, an unnecessary interference. Design and the management of design are practices apart. Finally, the fear of alienation from designers is compounded by the fear of a lukewarm reception from management colleagues: how seriously will they be taken by managers? Might they not continue to be considered as designers, or perhaps merely as dilettantes?

Surely, instead of transforming designers into design 'directors', and grafting them onto industrial organisations, a more sensible approach would be to make designers and managers better at their respective jobs, while increasing the sensitivity and understanding between them? Clearly, management eyes must be opened to the varying nature of design projects and the project management implications. Furthermore, the potential contribution design can make to the profitability of their organisations must be explained *and* demonstrated. This education has little to do with aesthetics and 'taste'; rather it must be concerned with the practicalities of the management of design projects and the building up of shared experiences. The objective should not be primarily to produce 'visual literates' but confident project managers who can use the talents of design specialists effectively. This approach offers the hope of improvements in the shorter term and, because the common ground and common endeavour are promoted, a much reduced potential for conflict. Furthermore, this approach is as relevant to companies which deal with design on a decentralised basis as it is to companies which control design centrally. Indeed it could be claimed that the approach is equally valuable to companies which have either weak or strong perceptions of design.

But how should companies go about upgrading the way they manage design projects? Consider the starting point: how do companies vary in their perceptions of design?

How perceptions of design vary from company to company

At one extreme, there are companies in which the perception of design as a coordinated set of activities with a tangible contribution to profitability is weak. Professional designers are not used at all. The need may never have been felt acutely enough, perhaps because design skills are not considered separate from other skills: an engineer is perceived to have requisite product design skills; a printer to have graphic design skills, and so on. Perhaps in the absence of any coordination, the use of many different specialists obscures the common ground and interrelationships between separate design activities within the company. Or perhaps designers are only associated with prestige projects, not bread-and-butter jobs. Designers may have been used in the past but without success; the results produced might have been judged to be technically deficient or the response from consumers lukewarm. Perhaps the consensus of opinion was that, though the designers produced interesting work, they did not fit into the organisation.

Frequently, a decentralised approach is adopted in these companies, with few constraints imposed from the centre: departments deal predominantly with their own design needs and little consultation takes place between departments. If design activities are sporadic, disparate and relatively unimportant, a case could be made for operating managers to continue handling the design requirements within their own areas independently, with coordination being introduced perhaps only insofar as one manager confers informally with others. In addition, if a clear corporate identity is not seen to be important, the case gains strength. In these circumstances, it could be argued that a centralised, separately identifiable design function is not necessary, Indeed, in the extreme case, all the work could be commissioned out.

As the volume, complexity and importance of design activity increases, the need to integrate the work experiences and operational interests of different departments grows. So too does the need for centralised control.

A progression may take place. Consultation between operating managers becomes more formal; perceptions of design grow clearer and more design activities are differentiated from other departmental activities. From common experiences, operating procedures are drawn up. The decentralised, 'laissez-faire' approach remains common, but designers are employed on some projects, though they are required to respond only to problems set and their work is tightly programmed. As the status of design rises, the work on one or two categories of design project is coordinated. Then the task of administering a few, then most, of the design activities is handed over to one person — perhaps first as a subsidiary responsibility, then as a major responsibility, and finally as a sole responsibility. Whereas limited design exercises can be administered relatively low down the management hierarchy, more sensitive, wider-ranging design projects require more senior personnel, and control would need to be exercised from higher up the hierarchy, if not the top. Consequently, as the importance of design activity increases, the appropriate status of the 'corporate design manager' rises.

Towards the opposite extreme of the spectrum there are companies where a professional approach to design pervades almost every activity, and it would be inconceivable for a product or any other output to be produced without the introduction of a designer early on in each project. In these companies, design represents a field of investment like many others in which profitability has justified the resources being used in establishing and maintaining the necessary quality standards. All design activities will be coordinated with the extensive use not only of operating procedures, but also of strict corporate design guidelines. Designers will frequently be given open briefs to seek out problem areas and to propose projects. Design contributions will feature prominently in company plans and senior managers will frequently champion this work.

Design 'responsibility' and design 'management'

At this point, it is worth restating the differences between the design project *management* role and that of design *responsibility*. Broadly, the project manager's role encompasses only the organisation and day-to-day administration of design projects. Though these individuals are often called 'design managers', they do not have the authority to set budgets for the projects they handle, nor do they have the final say. This authority rests with design responsibles — the controllers of investments and the final decision-makers in design projects. Whereas the design manager's principal role is to organise, the design responsible's principal role is to provide leadership. Of course, the roles of design manager and design responsible are not mutually exclusive: many design responsibles also manage design projects and many design managers are vested with responsibility for the projects they handle (see Chapter 5).

Responsibility for design activities need not be fixed at a particular level of a company. As the discussion has already revealed, analyses of such activities invariably lead to the realisation that certain projects can be dealt with low down an organisation while others require the participation of the board. The essential point in the efficient management of design activities is to assign individual projects to the appropriate levels in the management hierarchy. If projects are assigned too high up, executive time is wasted. Yet it is not uncommon to find senior managers dabbling with minor, routine questions of design. If projects are assigned too low down the hierarchy, then there is the danger that they may not be handled effectively.

Now the typing of design projects and the drawing up of design activity audits are topics for another discussion (see Chapter 2). But the project characteristics which should be considered in order to determine the level at which individual design projects ought to be assigned are summarised in Table 6.1.

Clearly, all projects ought to be analysed separately and projects with similar characteristics may well be assigned to different levels in different companies. Normally, as the characteristics of a project move across the Table from left to right, the importance of the project to the client organisation increases and, consequently, the appropriate levels of administration and responsibility rise. Projects with characteristics which lie predominantly to the left of the Table tend to be projects which can be executed relatively low down an organisation. Furthermore, the differences in status between manager and responsible can be small; there is rarely any justification in having senior executives actively assigned to such projects. Projects with characteristics which lie predominantly to the right of the Table tend to require the active participation of senior executives. Design responsibility ought to be vested at a high level though it may not always be necessary for the design manager to be of similar status, given that senior design responsibles are actively involved throughout.

Table 6.1 Factors affecting the nature of design problems

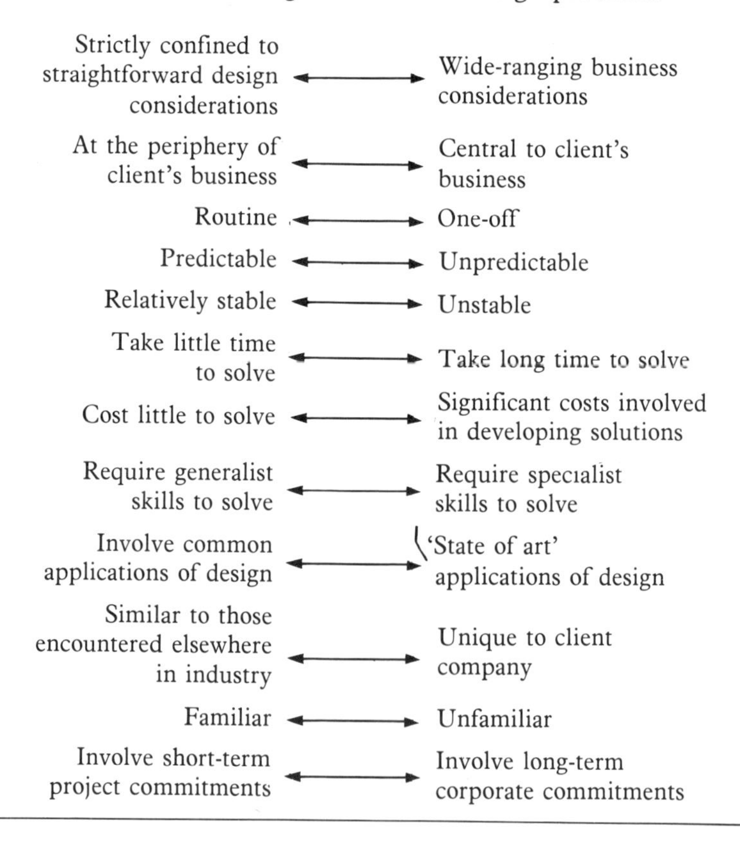

Strictly confined to straightforward design considerations	←→	Wide-ranging business considerations
At the periphery of client's business	←→	Central to client's business
Routine	←→	One-off
Predictable	←→	Unpredictable
Relatively stable	←→	Unstable
Take little time to solve	←→	Take long time to solve
Cost little to solve	←→	Significant costs involved in developing solutions
Require generalist skills to solve	←→	Require specialist skills to solve
Involve common applications of design	←→	'State of art' applications of design
Similar to those encountered elsewhere in industry	←→	Unique to client company
Familiar	←→	Unfamiliar
Involve short-term project commitments	←→	Involve long-term corporate commitments

Thus, interestingly, there may be a great difference in status between manager and responsible.

There are two points relating to design responsibility which do not vary from company to company. The first is that responsibility for any company's design activity can never be passed to an outsider; it *always* remains within the company. The second is that ultimate design responsibility always rests with the chief executive and *not* the company board — whether he acknowledges this responsibility or not, and whether he likes it or not. Unfortunately, relatively few chief executives take a serious interest in the design activities of their companies and, therefore, many are not in a position to discharge this responsibility effectively.

What does this kind of responsibility entail? To start with, the chief executive is accountable for all the design work undertaken by, or on behalf of, his company. It is he who provides the lead and backs the quality of designs produced. It is he who must ensure that design policies are translated into reality throughout the organisation and that the board is fully aware of the breadth of design activity within the organisation. These responsibilities may be discharged effectively with varying degrees of personal involvement.

Other aspects of this responsibility just cannot be discharged effectively without an active personal involvement in certain design activities. For example, the search for new product concepts (or service concepts, for that matter) must be steered by the chief executive. Similarly, design projects which are seen to be of particular significance to the company — such as corporate identity exercises — must also be steered by the chief executive. In neither of these examples is a truly corporate effort likely to result without the chief executive taking charge to moderate and harness the various departmental interests and energies.

So should the chief executive also be the corporate design manager? Where design activities are not formally under the control of named executives (as is the case in companies which adopt an extreme, decentralised approach to design problem-solving), the chief executive is also the corporate design manager by default. But in companies where the nature of design activity does not indicate a need for coordination, a corporate design manager will probably not be required — not at board level, anyway: several design managers can be assigned to handle the various projects under the watchful eye of the chief executive.

It should also be pointed out that, unlike the role of design responsible, the role of design manager need not always be held by a company employee. Indeed, if the truth be known, a high proportion of design projects are managed by designers on behalf of their clients. There is no reason why many companies should not seriously consider appointing an experienced outsider as corporate design manager.

Having carefully selected independent designers acting, in effect, as 'consultant design managers' can be a particularly useful solution where a part-time commitment is seen to be adequate. Many companies should find this an efficient way of keeping up with design developments and trends, as well as maintaining freshness in design work.

A framework of tasks for corporate design managers

Suppose, then, that a company wishes to create the position of corporate design manager, what tasks might fall within the scope of the job?

Clearly, these tasks will vary from company to company. Nevertheless a framework can be put forward in which the tasks are grouped into five areas. Before presenting the framework in full, a few words of introduction on each of these areas would be appropriate.

First, *planning:* two advantages of having a corporate design manager are that there ought to be greater opportunity for design investments to be planned ahead effectively and, consequently, for such investments to be allied more closely to the attainment of corporate objectives. The distillation of design objectives from corporate objectives, and the introduction of specific design contributions into corporate plans, are principal planning tasks.

Second, *monitoring:* the impact generated by design is not absolute, but varies with environmental circumstances. Thus the performance of company products must be monitored against those of competitors and market needs; likewise company images. The range of corporate design activities must also be monitored and the quality of design work constantly checked against specified standards.

Third, *organisation/coordination/control:* the majority of tasks in this area are directed towards ensuring that appropriate design problems are tackled — efficiently and effectively.

Fourth, *advisory/educational:* the generation of an environment of acceptance within the company for a more professional approach to design is often essential in any move towards higher design standards. Helping management to handle design projects more effectively and increasing the general understanding of the potential contribution of design to the profitability of the company are principal tasks in achieving this objective.

Finally, *reference/keeping up-to-date:* the collation of a company's design knowledge and experience into an easily accessible body of shared experience, and the maintenance of close contacts with designers and technical developments are the twin foundations of a useful reference service within a company. And the essence of a reference service is not merely to *enable* design work, but rather to *increase the productivity* of design resources.

The proposed framework of tasks for corporate design managers is given in Table 6.2.

Beware the panacea peddlers!

As in many other areas of management, what a company gets out of design management generally or a corporate design manager specifically will depend very much on what it puts in. A corporate design manager whose duties are set too narrowly and who receives little backing from the chief executive may well be more of a liability than no design manager at all. It does not always pay to start in a small way: for every company, there is a minimum effort required as a sound foundation for development. That minimum effort can only be gauged by examining an audit of corporate design activities.

Setting out the manager's duties too widely can also turn into a liability. Yet it is crucial to keep in mind that, in most companies, design pervades most activities. Therefore, if the boards of such companies are serious about developing more professional approaches to design problem-solving, they must be prepared to undertake progressively broader commitments across the spectrum of their activities. If that determination is not there, it may be better not to embark on any changes.

Design management, it would seem, is in danger of becoming a fashionable management concept, and where fashion takes a grip, common sense sometimes flies out the window. No doubt the panacea peddlers will try to promote design management and corporate design managers as they have many other concepts; however, we who serve in the design professions should be farsighted enough to issue a clear warning to the captains of industry: do *not* buy unless you have thought carefully why you need the product and how you will use it.

Table 6.2 Framework of tasks for corporate design managers

A Planning

1 To distil corporate design objectives out of the company's corporate objectives from which a corporate design programme can be formulated and introduced into long-term plans.

2 To propose and plan future design contributions to the whole range of organisational outputs, and hence corporate image.

3 To propose and plan further investments in design skills and facilities.

B Monitoring

1 To carry out regular audits of corporate design activities (annual or every two years).

2 To check the quality of all design work done within or for the company.

3 To monitor and evaluate investments in/expenditure on design.

4 To monitor the company's design standards and adjust where necessary.

5 To monitor individual output images, and to prepare audits of such images vis-à-vis the competition.

6 To monitor designs of competitive products, etc.

7 To monitor the company's image with target audiences, and compare with competitors' corporate images.

8 To monitor design trends and technical developments.

C Organisation/Coordination/Control

1 To diagnose problems that need to be tackled, and to ensure that the appropriate design problems *are* tackled.

2 To help set up design projects and, where necessary, to manage them.

3 To assist in the preparation of design project briefs and, where necessary, detailed project proposals for submission to relevant sanctioning bodies.

4 To bring together management and staff members required in any design project team.

5 To organise for the appointment and briefing of designers and other specialists.

6 To draw up work programmes for design projects.

7 To ensure that adequate resources are allocated to design projects, and that the allocation of resources is balanced according to priorities specified in the corporate design programme.

8 To coordinate the efforts of all involved in design projects.

9 To maintain the momentum of projects according to the various programmes set.

10 To ensure that all parties associated with design projects are kept informed of progress.

11 To prepare target audiences within or outside the organisation for the solutions generated from design projects. Where necessary, the bases for, and the operational implications of the design solutions should be explained.

12 To ensure that design solutions are implemented correctly and efficiently.

13 To apply a fresh, design-orientated mind to the solution of company problems.

14 To introduce designers/design consultants as fresh minds to be applied to the solution of company problems (perhaps as members of business teams).

15 Where appropriate, to build up the company's design expertise through the recruitment of in-house designers and the direction of the in-house design facility.

16 To perform a 'gatekeeper' role between designers, management and other staff.

17 To establish and ensure the maintenance of appropriate corporate design standards, given the company's corporate objectives.

18 To interpret design standards where necessary.

19 To resolve difficulties that may arise related to design aspects of any project undertaken, particularly where aesthetics are concerned.

20 To be responsible for managing the design of principal corporate outputs such as annual reports, etc.

D Advisory/Educational

1 To generate an accepting environment within the company for a more professional approach to design and for higher design standards.

2 To help managers and staff handle design projects more effectively by advising on all aspects of setting up and managing design projects — particularly on the diagnoses of design problems, the appointment of suitable designers, fees, and contractual matters.

3 To keep management and staff informed of the contribution designers can make to the solution of different types of problems and the attainment of corporate objectives.

4 To bring to the notice of management and staff relevant results of design projects and technical developments carried out elsewhere.

5 To keep management and staff informed of on-going design projects and the results of projects completed within the company — either through formal reports or informal newssheets.

6 To advise the board of directors on the company image required to achieve operating objectives and to maintain or enhance profitability.

7 To advise on and organise staff training in all aspects of design, and provide design inputs for other training courses.

E Reference/Keeping up-to-date

1 To collate the company's design knowledge and experience.

2 To act as a listening post for design ideas proposed throughout the company.

3 To maintain contact with designers/design teams and suppliers.

4 To keep abreast of design trends and technical developments.

5 To maintain contact with the principal professional design bodies and colleges.

6 To keep a register of designers/design teams and suppliers who could be used in different types of design project.

7 To provide an information service on all design matters.

7 Reflecting reality through design project documentation

It appears to be a commonly held view among managers that creativity ought not to be constrained through control. Unfortunately, when handling design projects, not only do they tend to be lax with the designers involved; they also fail to control the projects generally. To many such managers, design projects do not fall within the serious side of their duties. Consequently, strict control of such activities is not seen to be essential. Perhaps, too, design projects are considered light relief from weightier concerns of finance, production and general marketing. Thus to monitor design projects rigorously would be 'unfair'. Of course, where professional designers are involved, a high proportion of projects are not controlled by the client at all, but rather by the senior designer on the project.

Another commonly held view is that designers abhor administration and control and are, in several instances, quite unmanageable. It is indeed true that designers do begrudge time spent 'away from the drawing board' — administering projects, attending seemingly endless meetings, and so on. However, that generalisation belies the fact that competent designers do realise that the results and standards they seek would be largely unattainable without appropriate controls. Far from being unmanageable, these designers are often efficient administrators. Inevitably, there are designers who score high on creativity and technique but have neither the ability nor the temperament to take on more than a limited administrative role. Contrary to the widely held belief among designers, a competence in administration does not go hand in hand with a competence in design. In administration, a logical mind is crucial; in design, the ability to make effective associations — often arrived at by stepping deliberately outside logical analysis — is of far greater importance. Furthermore, a designer who organises his own work efficiently may not be as successful when organising the work of other designers, let alone the contribution of the client.

Managers who do not take design projects seriously and fail to organise them with care court disaster. Clearly the chances of generating effective solutions are substantially reduced, however competent the designers involved might be. Even if effective solutions are conceived, the projects may still represent poor returns on investment because of the inefficient way they were handled. The attitudes of managers towards design projects and designers add to the potential for failure. What many managers and designers fail to grasp is that, in a high proportion of design projects, the only common ground the designer and client share is when dealing with project administration. A manager may be 'visually illiterate' and have 'appalling taste', but he should understand administration — it is, after all, his professional territory. Similarly, a designer may know nothing of his client's business, but he will generally have a keen sense as to how a particular type of project ought to be organised in order to enable an effective solution to be conceived and implemented. There is little, if anything, which differentiates the professional designer's view of efficient project administration from that of professional managers. And there ought to be no problems with language: broadly speaking, the terms used are similar. There may be some divergence, however, where priorities are concerned. Normally a compromise is made between quality and standards achievable within the cost and time available. Frequently the extent of the compromise is influenced not so much by effective arguments on visual excellence or style, but by the more immediate and practical possibilities of project administration. Inevitably there has to be a trust between the parties that just so much, or even much more, can be delivered within a given set of circumstances. By working together through the formulation of a project work programme the true extent of these possibilities is often revealed, and trust is built up between the designer and his client.

Is a system of design project documentation always necessary?

Clearly, efficient project administration does not involve cumbersome documentation. It is possible to handle projects effectively without any documentation whatsoever — but there are risks attached to such a basic approach. No system of project documentation must be allowed to become an end in itself; at all times the purpose of such systems should be to assist those involved in the project to achieve the objectives set as efficiently and effectively as possible.

A *system* of project documentation will normally be most useful in companies which deal with many design projects concurrently, where projects need to be controlled continuously. Companies which handle design projects infrequently or projects which simply need detailed planning may see less advantages in operating such systems.

Project documentation concerns those who are actively involved in projects, those who are ultimately responsible for projects and, in some cases, those who will be affected by projects but are not involved other than by being consulted or being given information on certain aspects. Generally, those who will be affected by the progress or results of a project but are not consulted in any way will not need to be concerned with project documentation; their needs for information on what is going on should be catered for by means of internal memoranda or company newsletters.

A system of documentation must provide *records* of objectives, progress and achievement. It must also provide *evidence* of why and how a particular plan of action was taken. This information should enable the progress of a project to be assessed, and hence controlled. Such records and evidence are valuable references when projects are reviewed during their course or evaluated after completion. Should projects need to be repeated at a later date, perhaps as re-orders, these records will be principal *reminders*.

Project documentation can *facilitate communication* among members of project teams, and between project teams and others inside or outside the client companies. For example, a written brief is an ideal vehicle by which members of the project team and others in the client company can *look back* on the problem set within the context of the organisation and its markets, and *forward* to see how the proposed plan of action will contribute to the solution of that problem and any wider implications. At first-draft stage, a brief can prove to be invaluable as a generator of ideas and feedback, if only because it concentrates attention and can highlight shortcomings in the structuring of the problem and the project configuration.

It is worth looking more closely into the nature of communication between the members of project teams and others. Consider the case of a fairly important project with which the company board is involved. It may be that the board actually initiated the project by indicating concern for a particular problem area. Yet before work can proceed in earnest, specific objectives have to be set, a work programme drawn up and a budget indicated. The board must then sanction the project. Later on, the board will monitor progress — perhaps taking an active part when exceptional problems occur, when sensitive groundwork has to be cleared politically, or when solutions need to be championed in order to maintain the momentum of the project or to enhance the chances of a successful outcome. Towards the end of a project and in the following months, the board will evaluate the project, review the approach and standards adopted, draw conclusions and perhaps make recommendations with regard to future projects. The nature of the documents relating to requests for the sanctioning of projects, monitoring progress and reviews ought to be of particular interest to board members, for if they do not receive appropriate information and documentation, they will not be in a position to carry out their responsibilities. Understandably, the material the board receives need not include details of the day-to-day administration of projects, but rather summary information which (1) sets projects and their objectives within the context of the client company, as it is at present and in its future aspirations, and

(2) confirms progress or indicates actual or anticipated deviations from the set programme. Similarly, given that the board does not become actively involved in a project, the material which it issues will be in the nature of policy guidelines on operations and standards which would then be translated into action, either during the remainder of the project or with future projects.

With projects of lesser importance, this documentation may be the concern of an individual lower down the management hierarchy who is given responsibility for the project in that he has the authority to sanction the budget and has the final say on the strategy adopted and solutions proposed.

Within the project team, the documentation tends to relate to (1) background research into the project problem and the particular circumstances that give rise to it, (2) the formulation of the brief, the drawing up of the work programme, and calculation of the budget, (3) the formation of the project team and selection of designers, and (4) the day-to-day administration of the project covering the issue of orders, instructions to team members, the clarification of various issues, and so on.

Some design guidelines on project documentation systems

A number of guidelines could be put forward on the design of a system of project documentation:

1 It should involve the handling of a bare minimum of information and detail consistent with making suitably informed decisions. Redundant data or descriptions should neither be sought nor processed.
2 It should consist of the smallest number of items by which the necessary information and decisions can be communicated efficiently; that is, each item contains all the information required by the recipient, in a form which can be understood and used with the minimum of inconvenience. The existence and use of these items should be known to all connected with the project.
3 Given that specialists and managers from different backgrounds and seniority are involved in most design projects, the language used should be as simple as possible to ensure that messages are clear and are understood *as intended:* there is little room for jargon unless it is essential to efficiency in communication. In these instances, terms should be clearly defined and agreed at the start of projects.
4 The design of the documentation system should take into account the management style and other reporting systems within the client organisation. Ultimately, such systems are created to enable people to carry out their work more effectively; to minimise confusion and

discussion, they have to take their place alongside, and integrate with, existing management information systems. There is often a compromise struck between an increase in quality of decisions and the complexity of documentation. A complicated system which gives an accurate picture of the status of a project but is cumbersome to operate may create as many problems as a simple, easy-to-operate system which does not represent the reality of the project. Typically, the compromise is struck roughly at the point where what is perceived to be possible is balanced by what would be desirable. As the users' experience and confidence grow, perceptions of the possible and the desirable alter, and should gradually draw closer together.

5 The system should require little clerical and computing effort. The easier a system is to operate and the less time involved, the more likely it is to gain acceptance and the greater the attention paid to what it enables to be done rather than what it is. A system which is not used properly is almost as useless as no system at all. Ideally, systems of project documentation should be operated by those necessarily involved in design projects without any need for additional administrative staff.

What are the important items of project documentation?

Clearly no one system is likely to satisfy all companies. Therefore, how might an effective system of documentation be set up, and what would its components be? Experience suggests that the most important items of project documentation are the following:

For the client:

1 *The brief* through which the project problem is normally defined and a solution is specified. A brief should consist of:
(a) the business brief: that is, the definition of the business problem together with necessary details on the particular circumstances in the client organisation and markets;
(b) the design brief: that is, the definition of the design problem;
(c) the work programme broken down into stages; elements of work, who does what and who is responsible in the client organisation or elsewhere, resources allocated, deadlines, when work will be presented, how this work will be presented, and associated budgets.
   Where appropriate, a solution brief might also be prepared to provide guidance on unacceptable, inappropriate and/or anticipated solutions.

2 *The project proposal* – submitted cither to the company board for major projects, or to the manager responsible for the project in the

case of less important projects — relates to the initiation of projects. Different companies have different ways of putting forward projects for sanctioning. However, as a minimum, the information submitted to the board or responsible should include:
(a) the title of the project;
(b) the name of the department which is principally involved;
(c) a brief account of the background to the project, set within the context of the client organisation and its markets;
(d) the objectives set and benefits anticipated;
(e) other criteria against which the project will be judged for success;
(f) proposed starting date of project and anticipated date of completion;
(g) outline of project programme;
(h) outputs per stage;
(i) allocation of resources to project;
(j) total budget required now and at specified periods later on.

The actual brief formulated for the project could be attached as backup material to this submission.

When the project is sanctioned, it should be given a reference number and the priority accorded to it indicated. All material and documentation relating to that project should then bear that number.

Job numbers are generally given to complete projects. They might also be given to discrete parts of major projects which, for operational or administrative reasons, are better separated out. Furthermore, should the scope of a project become enlarged as work progresses, the additional involvement can (in certain circumstances) be treated as a new project and given a number of its own. By labelling different sections of work, there is a clear demonstration of the different components, hence the chargeable elements of work: a project which spawns several job numbers is unlikely to be charged at the original budget. Thus when a product is being redesigned, it is sometimes appropriate to consider the product design as a separate but related project to the design of the packaging. An exhibition stand designed for one show which is subsequently used almost unchanged in another show should be done through a project with a second job number. A graphic design project which starts out with the design of a letterheading, proceeds to the design of a whole range of stationery, and then to the design of promotional literature, should sensibly be filed under several job numbers. Perhaps giving a separate job number for each item would be overdoing it, but items related, either by the timing of orders received or by the nature or processing of their designs, could be grouped under common numbers. Thus several leaflets, each designed to promote a product in a range, might be listed under one job number.

Job numbers help to ensure that projects are adequately

differentiated from the administrative viewpoint: that all material and information relating to each project are channelled correctly to the relevant teams or individuals and are held in the appropriate files.

With projects which are seen to involve important investments and whose outcomes represent broad, long-term commitments on the part of the client company, sanctions to proceed will probably need to be sought at several stages. For example, in the case of a corporate identity project, the first proposal might deal with the formal initiation of the project, the allocation of resources towards the appointment of necessary specialists and research leading up to the formulation of a comprehensive brief. On completion of that stage, a second proposal would seek the sanctioning of work up to the formulation of a solution. When the solution recommended is approved, a third proposal would seek the sanctioning of a programme of implementation. With product design projects the equivalent stages might involve, say, consumer/retailer research, product conception/prototyping/testing, and tooling-up/production/marketing.

3 *The job file:* the job file is a central reference on the project within the client company. This file will contain all the formal documents on the project — such as the sanctioned proposal, the brief, work programme, summaries of research findings, minutes of meetings, directives, quotes and specifications received from suppliers, copies of procurement orders made out — as well as correspondence. This file should bear:
   (a) the project number and title;
   (b) the date the project was formally initiated;
   (c) the anticipated date of completion together with indications of crucial deadlines;
   (d) the project manager's name, and the name of the person (or body) to whom he reports;
   (e) the total budget allocated.

   Details should also be included of:

   (f) all the departments in the client organisation which will be formally involved in the project;
   (g) indications of the parts of the project concerned;
   (h) the timings of these involvements;
   (i) the person in charge of each department's contribution.

If in-house designers are assigned to the project, their names should be included. If an outside design group is appointed, the name, address and telephone number should be listed, together with the name of the individual in that group responsible for the project work. It may also be wise to take down the names of senior designers assigned to the team. Similar details of named suppliers appointed direct to the project should also be listed.

There are many ways by which job files can be organised: in straight chronological order, by stage of the project, by area of project, and so on. For most projects, one file is sufficient, though some companies may prefer to have one file for day-to-day administration and another for overall contractual arrangements. Where there are likely to be many visits away from the client company — such as site visits to new premises in the case of interior projects — it may be prudent to have two identical project files: one for use on site, the other to remain at all times within the client organisation in the department handling the project. In this way, should the site file be lost or damaged, the project documentation is not lost and a new site file can be copied.

Apart from a job file, a *job chart* might be drawn illustrating the progress of the project, week by week or month by month, depending on the complexity and the overall time span. Normally, such a chart would incorporate the following factors:

(a) time allocated against time actually spent (calendar and man-hours);
(b) estimated costs against actual costs incurred;
(c) elements of work completed against those planned per period.

The initial configuration of the progress chart would derive directly from the work programme drawn up for the project. As the project takes its course, more details can be recorded onto the chart. Broad objectives for the various stages would be enlarged into more specific objectives associated with particular tasks. As successive stages are completed, so the commitments for future stages become more concrete and are incorporated into the chart. When mapping out the future course of a project it is always worth while pencilling in as much detail as possible. However, a degree of certainty must attach to every detail incorporated: normally the value of the chart diminishes greatly beyond the next major point of uncertainty.

It must be stressed that job charts should include details of the client contribution to the project — specifically in ensuring the cooperation of staff members, in enabling design work to proceed smoothly, and in supporting the solutions accepted and implemented. These are not factors which designers can control, nor should they be expected so to do.

Clearly, some design projects will require the preparation of very detailed networks, in which case the known techniques, such as critical path analysis, should be used.

4 *Progress reports:* with lengthy projects, progress reports are essential aids to monitoring progress. The purpose of progress reports is principally to ensure that the board or the individual responsible is

* Note that a 'design responsible' is the person who authorises project budgets and has the final say with regard to the strategy and solutions adopted. He will not necessarily be concerned with the day-to-day management of design projects.

kept sufficiently in touch with the project for decisions which are referred up to them to be made effectively.* These rarely need to be bulky items: in many cases, a single typed sheet will be perfectly adequate. The following details should be covered:
(a) the project number and title;
(b) the name of the project manager and the department principally involved;
(c) the stage reached or work completed;
(d) changes in work programme (if any), and reasons for these;
(e) factors limiting completion or success of project;
(f) current staff and other involvements;
(g) the cumulative costs to date;
(h) variances in costs and/or time;
(i) further variances anticipated;
(j) likely effect on the outcome of the project;
(k) current anticipated total cost of the project at completion.

Some companies view progress reports only as formal vehicles by which projects are monitored. Others prefer to use such reports as the bases for formal reviews, in which case fuller details will almost certainly need to be included in them. Some companies will consider a design project to be complete when the proposed solution is implemented. Others will follow the solution through to see how target audiences react to it, how it is used, or how well it sells. In the latter approach, progress reports will continue to be produced perhaps months after design work has ceased. However, a clearer picture will be gained of the impact generated by the design project.

The last progress report ought to offer an overall review of the project together with evaluations not only of the outcome, but also of the way the project was handled. Because of the pressures under which managers and designers work, there is a constant temptation to 'drop' projects as soon as work has been completed, in order to concentrate on new and on-going projects. However, this temptation must be resisted until a proper review of the project has been completed, involving all parties. The failure to carry out such a review can, in many instances, rob designers and their clients of a potent learning experience. What has been learnt — both good and bad — must be highlighted so that, through articulation, it can become part of the shared experience within the client company, as well as between managers and designers.

Such reports could be produced at regular intervals or on completion of each stage of a project. Guidelines could also be set for reports to be produced when certain circumstances prevail — for example, when an important cost turns out to have been grossly underestimated or when priorities in the programme have to be altered radically.

There are advantages in a system which incorporates broadly fixed reporting times rather than highly variable times for individual projects, chosen either by the project manager or the company board. The former arrangement, being 'automatic', is less prone to postponement when pressures mount. With regular, formal reports there is a greater chance that data will be collected and presented properly to an agreed format. There should be less of a temptation to call hasty 'review' meetings at which disruptive, 'snap' decisions are made. Of course, it could be argued that major and rapid shifts in direction may not be possible with such 'rigid' systems. Yet it is quite possible for review sessions to be convened if the circumstances so demand.

For the designer or design group:

5 *Job card:* the principal project management references within a design group will be the job card and the job file. The principles which relate to job numbering as described in the client system of course apply equally to the designer's system. Such numbering ensures that projects are adequately differentiated from the administrative viewpoint; that all material and information relating to each project are channelled correctly to the relevant teams or individuals and are held in the appropriate files. The job card will carry outline details of the
   (a) design group and client company job numbers;
   (b) client company name;
   (c) job title;
   (d) date of confirmation of project;
   (e) anticipated date of completion or critical deadlines, if any;
   (f) budget agreed or method of charging;
   (g) principal client contact (that is, the project manager in the client company);
   (h) other contacts in client company;
   (i) designer in charge of project.

Job cards also contain information on orders placed and other expenses on projects. Details of dates of orders, what was ordered, who from, where from, the order numbers, the costs, and amounts charged to the client (if different) will all be recorded.

A time sheet is often incorporated onto the job card. On these, all who are involved in the project will enter brief details of the work they have done, the time taken, and date. Whether a design group charges the same rate or different rates for its various members should not influence the fact that each individual must record his time involvement; only in this way can accurate data of staff involvement be generated. Otherwise it will not be possible to differentiate between the commitments of partners/senior designers and juniors/suppliers, between the time spent on administration/client meetings and design/creative activities.

Finally, job cards should indicate when invoices have been made out, the amount invoiced, and when they were paid. In this way, the running profit of any project can be calculated on a weekly or monthly basis. At the end of the project, the data on the job card will enable the gross profit to be calculated: that is, the total amount invoiced and received minus all outgoings. The profitability of the project will emerge when this figure is related to the total time spent on it, and the resulting rate judged against the rate per hour normally charged by the design group, either generally or for the particular type of work involved in the project. However, *profitability* should not be confused with the *productivity* of projects. A project which is unprofitable may turn out to be highly productive, in the sense that it generates more work for the designers concerned, not just with their existing client but with others.

Copies of orders and invoices could all be attached to the job card which, in turn, is contained in a job file.

The job card system is of immense value to designers, though few exploit it to the full. In its most basic configuration, the use of such a system demonstrates an organised approach to project administration. The records of time spent, staff involvement and expenditure provide a vital reference and back-up to invoices. They also provide the bases for assessments of the efficiency with which work is being carried out (when actual time involvements and costs are compared with forecasts), and for forecasting future commitments to a project. By aggregating the data on all current job cards, an insight is gained into the future work load of the design group; where bottlenecks are likely to occur or where gaps may appear. With this knowledge, appropriate action can be taken in good time. Though crises cannot always be avoided, the analysis of jobs in this way should result in fewer crises and a substantial reduction in the amount of time devoted to fire-fighting. Being forewarned not only helps designers, but also enables them to prepare their clients where necessary. The analysis of job cards and time sheets of projects completed over a period of, say, twelve or twenty-four months may also provide an indication of how accurate estimates of cost and time were over that period. If any significant general bias is discovered either way, then future estimates might sensibly be adjusted by a similar factor. But care should be taken to ensure that there is no explanation other than a general error in forecasting.

6 *The job file:* the job files compiled within design groups are very similar to those of client companies. However, the details contained on job cards may not all be repeated in the job files.

The job file will bear the following:

(a) the project number and title;

(b) the names and positions of all relevant contacts in the client company;

(c) the names, addresses and telephone numbers of all special suppliers who will be involved in the project.

The most important contents of the file will be the letter of appointment, the brief and the programme of work. The other contents will include all other matter relating to the project: minutes of meetings, quotes, specifications and back-up material received from suppliers, copies of procurement orders, photographic references relating to design material, samples (if these are bulky they may perhaps be contained in a separate box file), as well as correspondence.

With projects where many 'technical' drawings are produced — such as projects concerned with the design of signage schemes, interiors, exhibition stands and products — then a 'drawing file' is also a necessary component of the documentation system. This file will normally hold the master drawings or negatives, copies of which can be ordered as required. Each drawing will need to bear the project number as well as an individual drawing number and title for specific identification. Both the job and drawing files should hold an index of all drawings produced in the course of the project for quick reference. As soon as a drawing has been completed and approved, it should be entered into the 'drawing file', with the details and date recorded in the index. This index should also contain details of numbers of copies of all the drawings issued and to whom — data which will be transferred to the job card for costing/invoicing purposes.

Again, more than one job file may be compiled for administrative or security reasons.

Wanted: adequate information in 'unfamiliar' management territory

Managers perceive one of the principal difficulties encountered when handling design projects to be the significant lack of information at the start of such projects. Experience also suggests that managers are frequently no wiser *after* the completion of design projects.

One reason why this is so is that few managers attempt to evaluate what occurs during the course of such projects. Another reason is that if they were to try, many would not get very far because adequate records are not available to them.

There should be little surprise, therefore, that many managers look upon design projects as unfamiliar territory, even though *all* outputs produced by organisations are designed — effectively or otherwise.

Clearly designers do not always maintain adequate records of the work they undertake. Furthermore, many do not see it as being in their interest to explain their work to clients.

Without mutual understanding and the building up of common experience, confidence will not be generated. And without confidence, there is really little chance that the design talent of this country will be harnessed effectively by management.

So, between them, designers and their clients conspire to ensure that the contribution effective design can make to the profitability of business enterprises and to the quality of life is rarely examined. Improving the quality of project documentation will not, in itself, overcome these drawbacks. But such action should at least help improve the efficiency with which design projects are processed as well as open a few managers' and designers' eyes. This, surely, is a worthwhile start.

8 Counting the costs, but losing sight of the benefits

To date, the industrial design profession has failed to grasp the significance of evaluation in promoting the contribution effective design makes to economic activity and to the quality of life: spokesmen for the profession have preferred to rely instead on rather pious, largely unsubstantiated exhortations. The low status of design relative to other professions is a clear demonstration that, in practice, the 'acceptable face of design' has not impressed management. Worse still, there is precious little evidence that management is learning anything from the salesmen of 'good design'.

There are two major reasons why a serious study of design project evaluation should be undertaken in the next few years. The first is that, in the long term, there is no such thing as effective management without evaluation. If design resources are not managed effectively, it is difficult to visualise how design standards might be raised. The second is that, without understanding, there cannot be effective management. Through rigorous evaluation, management understanding of design and design projects will increase, and this must have a favourable effect on the way management invests in design and tackles design projects.

Are design projects unsuitable for evaluation?

What is so special about design and design projects that has led to the belief that they are unsuitable for evaluation? For both designers and clients tend to hold this belief. Indeed, few designers agree that evaluation could be of any value. By contrast, clients are generally quick to realise the potential benefits through evaluation. Nevertheless, attempts at evaluation are rare. Several reasons can be cited for this.

It is still widely held that 'things artistic' do not lend themselves to 'number-crunching' evaluations. There are strong protests when art, considered by many to be a vital civilising force, is reduced by the rich and large organisations to just another field of investment. It will be argued that quantifiable returns should not be sought from art, yet the essential need to support the arts financially will be stressed. Strong protests will be heard, too, when what appear to be jokes are made through the medium of art — in the form, perhaps, of piles of bricks or cans of soup. In general, we like our art to be good and clean, accessible yet full of wonder and mystery.

The links between perceptions of art and design are very strong; for many, there is no easy distinction between the two. Though the development of the ability to design was one of man's earliest achievements, the recognition of a distinct profession of 'designer' is only a recent development. Previously, we had artists and artisan craftsmen: the training of these 'designers' was steeped in art and the crafts. Thus the concept of 'goodness' in art — which is sometimes obvious only to an enlightened few — was transferred to design. For years, the skills of designers have been promoted principally through their alleged capacity to produce 'good design' — a quality which those not trained professionally, and thus largely unenlightened, could hardly expect to achieve, despite the fact that some of the most memorable designs have been produced by people with no formal design training. Not surprisingly, selling 'good design' has been an uphill struggle, partly because what constitutes 'good design' has never been adequately defined, partly because of man's innate capacity to design, and partly because the pinnacle of achievement in design is often manifest in design solutions so elegant in their simplicity that they appear to be within the competence of us all. How often do we exclaim, 'Even I could have done that!' when a more honest reaction might be, 'How refreshingly direct and obvious!'

Designers who are good at their work will be described as possessing a 'feel' for what is right; 'taste' is another word commonly used which, like 'good design', has never been satisfactorily defined. Successful designers may also be described as having 'flair', a quality which, boiled down, amounts to an ethereal propensity to 'gut feel' successfully. Flair is maintained to be a quality which cannot be evaluated because there is nothing logical or teachable about it. Consequently it is suggested that those things that can be evaluated in design projects will not include the designer's most distinguishing contribution to the satisfaction of needs and the solution of problems.

The subjective, intangible qualities which are perceived to constitute 'good design' are difficult enough to describe and discuss; they are even more difficult to write about. Indeed, little of substance has been written about design evaluation; almost nothing has been written about design *project* evaluation.

Design projects are often looked upon as investments in prestige — an attitude conducive to the development of a certain laxity of control. In those instances where top managers initiate design investments, lower management might perceive these to be pet projects and decide that rigorous evaluation would be 'indelicate'.

The fact that most managers have no clear conception of what designers do or how they go about their work adds to the unease of having to deal with design requirements, which can be highly subjective. The result is that few managers communicate effectively with their designers: many a design project is launched with ill-defined objectives, with a client and designer who have neither grasped the problem nor agreed on the scope of the project. Managers will rationalise such shortcomings by suggesting that to define a problem too narrowly and to be too strict when dealing with designers muzzles their creativity: many projects are kept wide open 'to see what the designer comes up with'. Managers will also explain that designers should not be overloaded with information; that they know what information they need to solve problems and will ask for more as and when necessary. Unfortunately, designers will normally have little knowledge of a client's business before they start work on a particular project. Without such knowledge, relatively few will know exactly what questions to ask in order to define a problem adequately. Frequently designers will not wish to probe too deeply, as their ignorance might be revealed. In any case, clients are not always willing to divulge too many details before a project gets under way, and can become very irritated by pointed questions during exploratory discussions. Thus many designers and their clients prefer to get started on projects before covering the detailed groundwork which invariably constitutes the foundation to a successful outcome. Without a clear, negotiated brief, the chances of undertaking any sensible project evaluation are very slim indeed.

Just as communication between client and designer can be ineffective, so too can the communication between designer and other members of the client organisation. Even the dialogue between a client and his colleagues may leave much to be desired. Typically, proper records are not kept on the progress of design projects: correspondence and project documentation tend to be sparse except when difficulties arise, particularly between designer and client. What written material is generated during the project is rarely kept for long, if at all, after the project is completed. Thus, unless evaluation takes place as a project progresses or shortly after completion, no documentation may be available to aid evaluations. What records exist will probably give an unfair and incomplete picture of progress and achievement.

When designers and their clients are perceived to speak different languages, when they understand little of each other's work and attitudes, and when no universally applicable system of evaluation is available, then attempts at evaluation are likely to be complex, lengthy and incomplete. Many designers already begrudge the time spent administering projects; they fear that evaluation will make equal demands on time which should otherwise be devoted to the creative process.

Both designers and managers can be inconsistent and biased in their assessments of projects. Important and unfamiliar areas will be glanced over or ignored, while relatively trivial but familiar areas are examined in detail. Highly subjective, snap assessments will be made in certain areas while quantitative assessments will be laboured in others.

It is not uncommon for designers to be blamed for all difficulties and failures encountered during the course of design projects, so rarely is the client contribution scrutinised. With most projects, the client involvement is neither incorporated formally into the work programme nor costed; 'officially' there may be nothing to evaluate. Nevertheless, the client contribution (or lack of it) often has a critical influence on the outcome of a project.

Many designers also feel threatened by evaluation, for they concede that with rigorous evaluation will come greater management understanding of their work. They suggest that a sufficient number of clients already get excessively involved in design projects; greater understanding will lead to an increase in interference, especially during the design process. In these circumstances, the influence a designer could exert on a client will diminish significantly. Clients, too, have grounds for unease about evaluation: if the client involvement in design projects currently receives only cursory attention, with rigorous evaluation any shortcomings on the client's part should be revealed.

Of course, there are so many different kinds of design project, and so many factors influence the outcomes of design projects — several of which are beyond the control of designer and client — that assessments of contributions and responsibility may be particularly difficult, if not impossible. Thus it can be argued, as designers often do, that evaluation can never be complete or accurate, and will almost certainly be counterproductive when undertaken by those who have neither sympathy for, nor sufficient understanding of, design matters.

However, the fact that design is a creative specialism in no way implies that design projects do not lend themselves to evaluation. In reality, design projects are similar to most other types of business project. In the hands of properly briefed professionals, design problems are no less structured than management problems; indeed, design decisions require no fewer skills and involve no more subjectivity than management decisions. The misconceptions concerning the nature of design and the potential contribution of designers are clearly deterrents to design project evaluation. Yet they give rise to project management behaviour which adds further deterrents to evaluation. There is ultimately no reason why investments in design should be appraised by any different standards than other, more common types of investment. Indeed, if such investments are to become frequent, it is *essential* that they be treated with equal rigour.

Basic concepts of evaluation

If we were to ask the question 'On what basis should design projects be evaluated?' what answers might we expect? Would designers answer differently from their clients? Over the past two years, this question has formed the basis of discussions with businessmen and designers in the United Kingdom and North America. There have been two striking similarities in the responses received. The first was an almost universal admission that the question had not been considered before. The second was that, when pressed, respondents tended to put forward only one suggestion: that the quality of solution created should form the basis for the evaluation of design projects.

Clearly to many the word 'evaluation' conjures up the image of a client attempting to assess the quality of solutions generated and the performance of designers. However, these aspects provide only a part of the picture, for it should be emphasised that in design project evaluation our interest encompasses the evaluation of the progress of whole projects, not merely the evaluation of outcomes. It is interesting to note that this image of evaluation is quite different from the practice of evaluation with other business, and particularly R & D, projects. In those cases, evaluation often occurs before the start of projects as part of a viability/selection procedure.

To evaluate is to establish what is put into a project (the *inputs*), what is derived from it (the *outputs*), and whether the outputs justify the inputs. Thus we talk of the costs of, and investments in, design projects that lead to benefits. Costs do not derive only from the inputs of resources at the start of projects; inputs can be made throughout the span of a project, and there are also costs associated with outputs — for

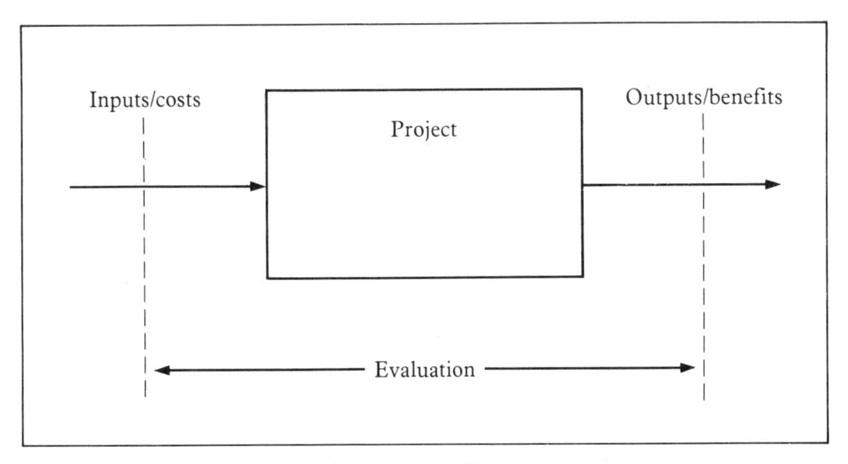

Figure 8.1 *Basic concept of project evaluation*

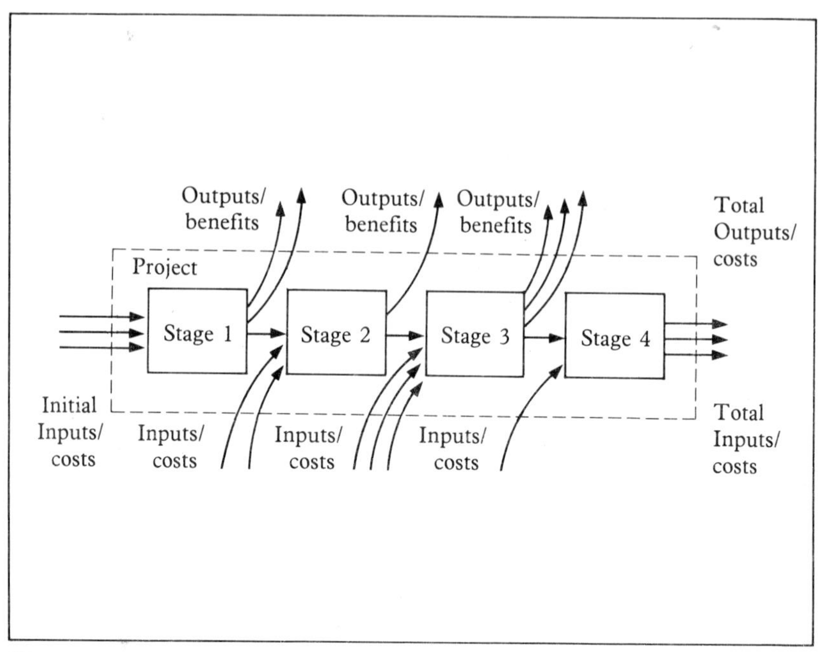

Figure 8.2 *Costs and benefits can occur throughout a project*

example, ineffective solutions cost money. There may also be significant costs in *not* undertaking a necessary project. Similarly, benefits need not derive only with end-products, after completion of projects; they can occur throughout a project, and certain benefits can be enjoyed while a project progresses. For example, investing more time and money in diagnosing a problem correctly and formulating an accurate brief are benefits which occur early on in a project and are enjoyed for the remainder of the project. As with other business project evaluations, expressions of costs and benefits need not be confined to financial statements.

Thus, in evaluation, *pre-project* circumstances are defined which are then compared with corresponding *post-project* circumstances. In order to make sensible assessments of the outcomes of projects, it is necessary to determine what was involved in transforming the pre-project circumstances into the post-project circumstances. The sorts of questions project evaluation should help to answer are: given the original client circumstances and the resources allocated to a project, were the results achieved satisfactory? Could the outcome have been forecast? Or alternatively, given the results achieved and the history of the project, do these tell us anything about the resources that should have been allocated to the project, or the way in which the project might have been organised? Design projects are usually set up because of perceived problems and needs. However, we are rarely interested only in whether or not a need was satisfied or a problem solved. We seek to establish, for example, how the problem was defined, what constraints were placed on the form of its solution and the manner in which it could be tackled, how effectively it was solved and the wider implications of implementing the solution. In other words, we seek to establish whether sufficient opportunity was given to solve a problem effectively, whether the opportunity was properly exploited, and whether the project was progressed efficiently.

Design project evaluation is not about making design decisions easier for managers; rather it is concerned with helping managers to make design decisions more professionally. Evaluation should provide the kind of detailed hindsight from which guidelines on effective design management practice might be distilled and confirmed. By broadening outlook and increasing knowledge, design project evaluation should help build a manager's confidence in his ability to handle design projects effectively and in the contribution designers make to the solution of his problems. This, in turn, should increase the willingness of management to invest further in design and to take greater risks during the course of design projects, albeit on the basis of better-informed decisions. In sum, design project evaluation is no panacea for incompetent management, and will be of little use to managers who do not seek to achieve high standards within their companies.

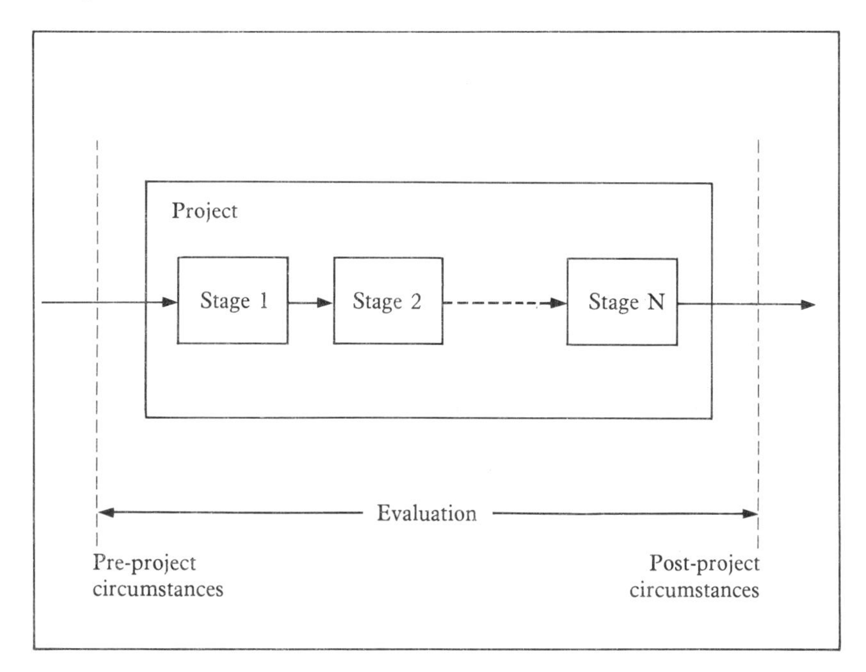

Figure 8.3 *Pre-project and post-project circumstances*

Different bases of project evaluation

Several important questions need to be answered on design project evaluation. What exactly do we need to evaluate? Do different types of project require different approaches to evaluation? How should evaluations be carried out? And, when might such evaluations be undertaken?

Design projects vary greatly in nature. A product design project is not the same as an exhibition design project; an interior design project is different from one which involves only graphic design.

However, it is not just in the 'mechanics' of these categories of project that differences occur: major variations can result, within categories, from the way they are handled by client and designer. Thus a project in which the client briefs the designer comprehensively and supervises his work closely is a very different kind of project from one for which the client does not formulate a proper brief and demonstrates little interest in its progress. Furthermore, a project in which the client dictates the solution to the designer is, by nature, different from one in which the client refrains from providing a brief with an implied solution and keeps an open mind as to the types of solution which would be acceptable.

The nature of a design project also influences the mix of skills required to tackle it: a mix which varies, not only between one project and another, but also within the lifespan of a single project. Broadly, these skills have been differentiated as creative, technical, business, administrative, and interpersonal.

The differences between these types of project relate to the point at which the designer is introduced and their scope, the division of work between client and designer (hence the influence each has, and the contribution each makes), and the consequent responsibility which can be apportioned for the outcome.

Despite these important differences, all design projects proceed through similar stages which provide part of the evaluation framework we seek.

A design project is normally set up because there is a perceived problem or need. The project problem is diagnosed and a brief formulated on the basis of which a solution is conceived. When the concept of solution is approved, it is designed in detail, implemented, and put to use, as a result of which an impact is generated.

Two fundamental points arise from this sequence. First, design projects should be organised to proceed through these stages smoothly and efficiently. Second, the fit between stages must be close. That is, the project problem should be diagnosed correctly; the brief should reflect the project problem accurately; the solution conceived should answer the brief effectively, and so on. Without a close fit between stages, a project will be increasingly deflected away from the basic objective: the solution of the problem which gave rise to the project.

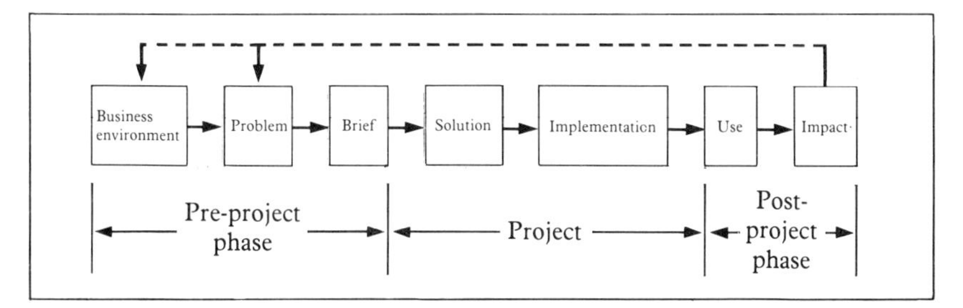

Figure 8.4 *Common stages in progress of design projects*

And without efficient links between stages, a project ceases to be a single coordinated exercise, and breaks up into several sub-projects. Such fragmentation of design projects is, unfortunately, common and has to be taken into account during evaluation.

If evaluation is to be a useful operational and planning tool, it must help determine *when* fragmentation occurs, as well as the causes of fragmentation. For example, it is insufficient to state that an inappropriate solution was conceived in answer to the brief. Was the failure due to the designer's inadequate creative skills, or could it be that the client was so disorganised that no designer, however creative, could have come up with a better solution? Had the designer failed to grasp important operational considerations, or was the brief too loose? Were the designer and client able to work harmoniously together? Answering such questions helps determine whether the outcomes resulted from creative/technical, administrative/business, or interpersonal failures – and should provide pointers for remedial action.

This sequence of stages can also clarify some of the implications of using 'quality of solution' as a basis for evaluation. A design solution starts out as a concept visualised in the mind's eye of its creator. This is translated into physical reality during the implementation stage. The appearance and physical construction of design solutions are the most accessible properties by which their quality is evaluated. Understandably, a fair number of designers and businessmen mentioned only these two fundamentals when questioned on indicators of quality.

Yet appearance and physical construction are rarely ends in themselves, but are means to ends, as shown by the sequence of stages. Designs have to function effectively in a prescribed manner, over a period of time; they should generate a planned impact, and so on. Thus, no assessment of quality of solution is complete without consideration of the way the solution performs and the impact it generates.

As all product manufacturers will attest, products do not necesssarily function efficiently (if at all) unless they are used correctly. Similarly, design solutions may not generate planned impacts unless implemented efficiently and used appropriately. We should not be surprised if a brilliantly designed product is a financial flop because it is neither produced efficiently nor marketed effectively. Our shops abound with well-designed merchandise, packaged and displayed indifferently — when available. We have grown accustomed to receiving literature which attracts us visually, but whose poor copy detracts from the products and services being promoted. Clearly, not all these factors are within the influence or control of the designer: neither is it necessarily desirable that they should be. Yet they do contribute to the tangible outcomes of design projects.

Thus, the sequence of stages gives us several bases for project evaluation. Ideally a project should be evaluated by the impact its outcome generates on the project problem. This is the acid test.

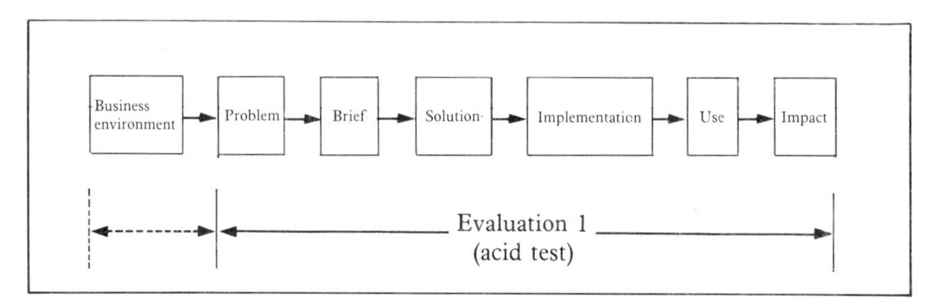

Figure 8.5 *Evaluation of impact generated by implemented solution on problem (the 'acid test')*

Nevertheless, it should not be forgotten that problems arise out of particular business circumstances, either within the client organisation or in the wider environment. As problems tend to be closely linked with business environments, no assessment of impact on a problem would be complete without some analysis of impact on the business environment. Clearly, it is no good solving one problem while creating other intolerable problems for the client. The use of the adjective 'intolerable' is deliberate. Design projects, when handled professionally and efficiently, can raise more problems than they solve initially. Luckily, the majority of these problems can reasonably be termed 'tolerable'. For what the client learns in the process tends to raise not only his sensitivity to his organisation's internal and external circumstances, but also his confidence in being able to influence and alter these — whether through effective design or otherwise. The learning experience represents one of the most powerful benefits derived. For example, it may be that, following the redesign of one product, a client realises the potential for improving other products or design standards generally within his organisation. Thus the impact, and consequently the benefits, of a project may extend well beyond the confines of the stated problem. It is surely a prime objective of professional designers to build up their clients' knowledge and confidence sufficiently for them to seek such productivity wherever possible from design projects.

Should the project become fragmented as it progresses, the sequence of stages still indicates how the project might be evaluated. There are two common areas of fragmentation in design projects. The first occurs at the start: the project problem is diagnosed incorrectly and an inaccurate brief drawn up. In these cases it is only reasonable to judge the impact of the solution against the brief provided. If the client drew up the brief and the designer had no influence over its content, then the responsibility for an ineffective project outcome lies squarely on the client's shoulders. On those occasions when the designer is the prime mover in the diagnosis of the problem and the formulation of the brief, then some burden of responsibility must be transferred to the designer, though the client retains ultimate responsibility for the failure. However, it is important to stress that, in both instances, the fragmentation was caused by a business/administrative failure, not by a creative/technical failure, even though creativity may play a part in the formulation of effective project briefs.

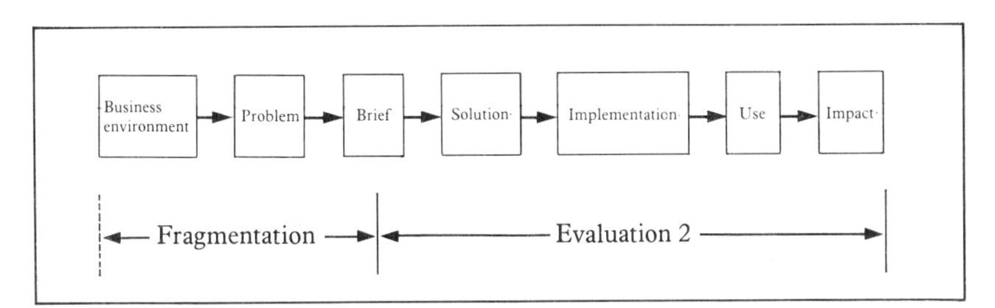

Figure 8.6 *Assessment of impact generated through a design project, given the brief*

The second common area of design project fragmentation occurs towards the end of projects: though the designer puts forward an effective solution, it is neither implemented properly nor used rigorously as prescribed. Perhaps the solution is implemented incorrectly by the client or another agent: specifications for colours, the use of materials and processes, or production tolerances are altered without the knowledge or approval of the designer. If the designer has no control over implementation or usage, then the consequences can hardly be blamed on him. If a design solution is not used correctly or effectively by the client, then again the designer should not be blamed for the consequences. A product which is not marketed properly is unlikely to sell well, however effective the design. A brochure which is sent out to the wrong audience is unlikely to generate the response aimed for. An exhibition stand which is inadequately staffed and is not tidied up regularly is unlikely to draw anything like the anticipated number of enquiries. Thus the contribution the designer makes to the project should strictly be judged only on the basis of the solution proposed, given the brief supplied. This is the most limited and immediate basis of evaluation. And as the respondents described at the beginning of this paper indicate, it is probably the basis used most frequently by clients and designers alike.

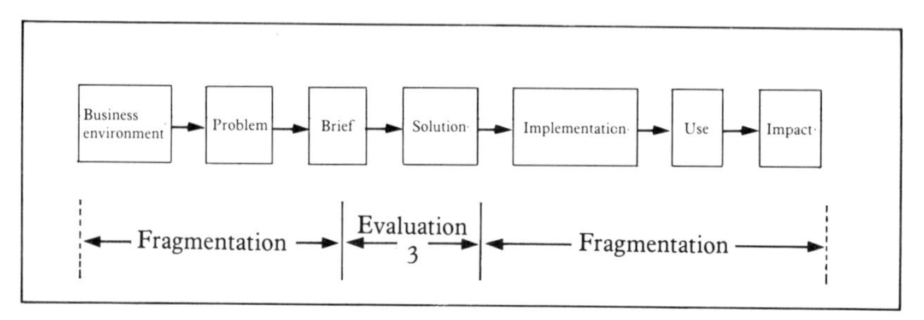

Figure 8.7 *Evaluation of design solution, given the brief*

What factors need to be evaluated?

The varying scopes of evaluation indicated in these sets of circumstances provide us with the three fundamental bases of design project evaluation. They also provide insights into the factors that need to be included in any attempts at rigorous and comprehensive evaluation. These will now be discussed in greater detail.

A solution needs to be assessed on several planes. Obviously the *appearance and physical construction* of a solution is of paramount importance: the forms, colours and materials used, and the ingenuity and care with which these are combined.

The *performance* of a solution is also an obvious consideration: what it does, to what extent and how efficiently it achieves the objectives set, will need to be determined. However, before a sensible assessment of return on investment can be reached, in the longer term, it will also be necessary to determine how the solution performs in changing circumstances which, perhaps, could not have been foreseen, and does not create unfortunate side-effects.

The *productivity* of solutions is a further consideration. Have more problems been solved than anticipated with the one solution? Has the impact of the solution been greater than hoped for: that is, have benefits emerged in unexpected areas (perhaps through the creation of new demands rather than just satisfying existing demand)? Or have benefits been enjoyed for longer than anticipated?

Finally, solutions can be analysed from the *project management* viewpoint. For example, is the solution easy to implement? Are the costs involved in implementation easy to justify? And so on.

The *designer's performance* must also be assessed from several angles.

It is often taken for granted that the designer's most important contribution derives from his ability to create effective solutions. Yet with many design projects, the development of such solutions would be virtually impossible, were the designer incapable of organising efficiently to undertake the creative and implementation work. As effective designs do not necessarily sell themselves either to clients or in the marketplace, it can also be argued that the designer's principal contribution is in creating the environment that will accept the most appropriate design solution, a solution which may not be the one sought by the client. Many designers can create outstanding solutions; not all will be able to convince their clients to accept such solutions. Thus the designer's creative, technical, administrative, business, and interpersonal skills and performance all come under scrutiny during evaluation.

Interpersonal skills are frequently the only foundation to the vitally important designer/client relationship. Depending on the kind of relationship established, doors are opened or closed to designers. Projects may succeed or fail for no other reason than that the client liked the designer and got on well with him, or that he took a disliking to the designer and did not cooperate.

When assessing the *contribution made by the client* to the project, again several areas must be investigated. What part did the client play in diagnosing the problem and formulating the brief? Was the

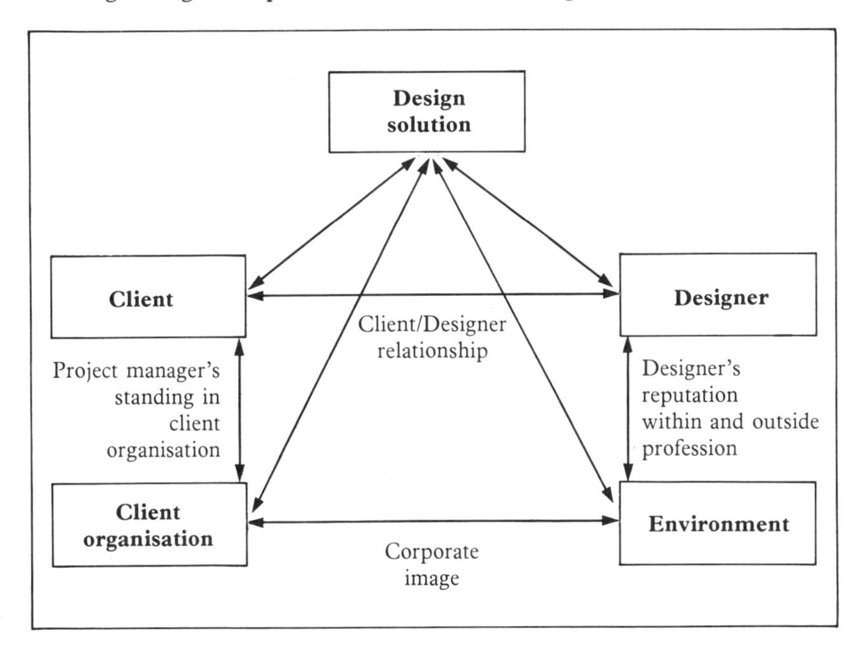

Figure 8.8 *Factors to be considered during the comprehensive evaluation of design projects*

diagnosis correct and was the brief accurate and clear? Was the right designer appointed to the project? Did the client ensure that the designer got the necessary cooperation and support from within his organisation? Did he provide sufficient information and make decisions when required to do so? Was the client organisation adequately prepared for the presentation and implementation of the solution proposed?

In analysing the outcome of a design project, an assessment of the *standing of the client (or project manager) within his organisation* may be critical. Has the project reinforced the client's point of view? Does the solution proposed raise few criticisms within his organisation? Is it thus easy to promote? Does the project successfully break new ground for the client organisation without creating any headaches? The answers to such points provide indicators of costs and benefits relating to the client *within* his organisation.

Last but not least, it is obviously necessary to evaluate the *impact a design project has on the client organisation in relation to its environment.* Has the solution succeeded in the marketplace? Does it establish the client organisation favourably apart from its competitors? And so on.

These component factors in design project evaluation and the relationships between them can be depicted in a pyramid, as a convenient, summary reference.

When should evaluation take place?

When should evaluation take place? The progress of projects from the problem and business environment through to the impact generated by the solution implemented suggests four principal stages at which evaluations ought to be considered.

The first of these is at the brief stage. At this point, it is important to determine the 'fit' between problem and brief. Does the brief reflect the problem accurately? Is there adequate information about the business environment included in the brief? What resources were put into the formulation of the brief — either by the client organisation, the designers, or other agencies? Do all parties involved in the project understand the problem and the objectives set for the project? Have the brief and all underlying commitments been formally agreed by all principal parties?

The second stage at which evaluation will take place is when design proposals are presented to the client. At this point, the fits between the solution proposed and the brief, problem and business environment will be examined. Do the proposals answer the brief? How well will the solution function? Are the proposals as expected? If not, how are the client organisation, retailer and customers likely to react? Can acceptance of the proposals be justified easily — financially or otherwise? Can the proposals be implemented efficiently without overstretching the client organisation? Was the right designer

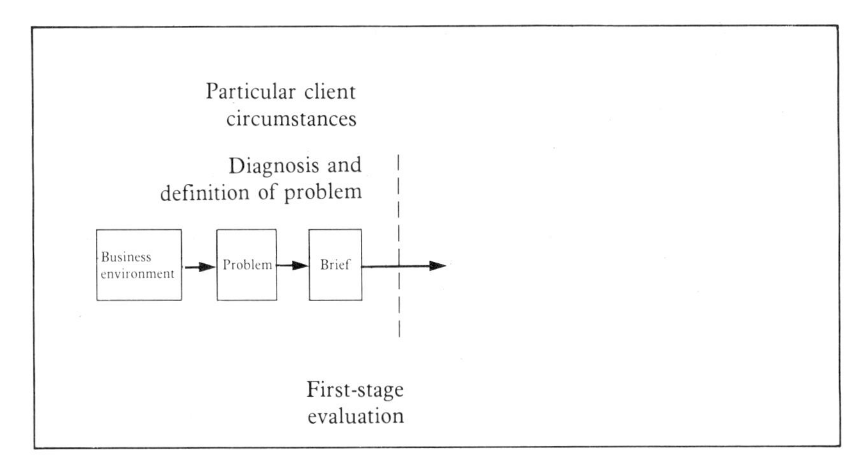

Figure 8.9 *First-stage evaluation of 'fit' between problem and brief*

appointed? What support did the designer get from the client? How efficiently was the project handled? Did designer and client work well together? It is at this stage that an immediate reaction is triggered off as to the progress of the project. Feelings of dissatisfaction which might have remained unexpressed pending sight of the proposals may be unleashed if proposals put forward do not live up to expectations. Alternatively, if the proposals far surpass expectations, there is the equal danger that project control slackens or that the scope of the project is expanded unreasonably. Maintaining an appropriate degree of objectivity and balance in these circumstances is one of the severe

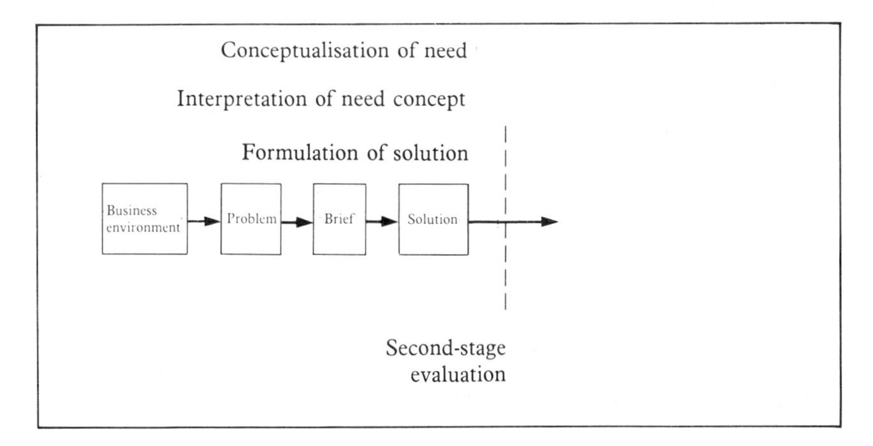

Figure 8.10 *Second-stage of evaluation involving assessment of 'fit' between solution and brief, solution and problem, and solution and business environment*

tests design project managers will encounter.

The third stage at which evaluation will be attempted is immediately after a solution has been implemented and put to use. At this point, the principal concerns will be to determine the fits between the solution as proposed and as implemented, between the solution and the use to which it is being put, the actual use being made of the solution and that which was anticipated at the brief stage, and finally the reactions from the business environment. This stage normally offers a short-term reaction to the outcome of the project.

The last stage of evaluation is the most difficult and complex of the four. It concerns the assessment of the long-term impact generated by

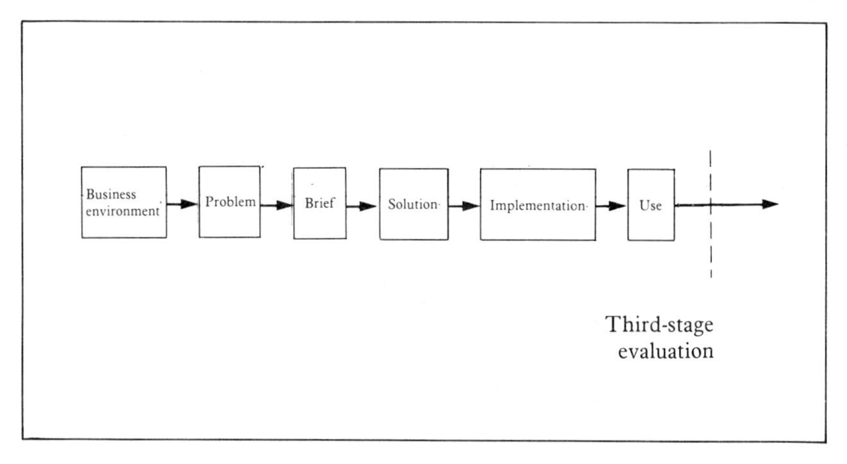

Figure 8.11 *Third-stage of evaluation involving assessment of 'fit' between design proposed and the way it is implemented and used*

the project on the project problem and the business environment which gave rise to it: the acid test referred to earlier. The question is: when, after completion of a project, should this evaluation take place? Is there a fixed guideline on this? Clearly there is not, for with some projects the timing will be specific and final, while with others it will be less definite and could be undertaken over a period of time. For a project set up to achieve a particular result at a specified time, the impact generated ought to be evaluated at the time specified or immediately afterwards. For projects which are set up to achieve specified results at particular times but are also anticipated to have extended impacts beyond those times, more than one evaluation is indicated. Take the example of an exhibition stand design. The first evaluation must surely take place straight after the exhibition closes. Comments on the stand design — and, more importantly, enquiries — may flow in for many months, right up to the following exhibition in which the company participates. It could be argued that the final evaluation on that project should take place just before, and as a foundation to, the follow-up exhibition project.

Determining costs and benefits

What are the practical difficulties encountered when determining the costs and benefits of design projects?

Briefly, costs tend to be easy to identify and quantify. Normally, they arise from expected and familiar sources, and it is relatively easy to anticipate when they will be incurred. Most costs are incurred during the lifespan of a project; few, if any, are foreseen after the completion of a project.

Costs are seen to be 'real' and tangible because they frequently have to be borne in the present or the immediate future, and because everyone understands the meaning of paying out money. When not related to the benefits sought, they often appear frighteningly large. The recording and analysis of costs lie within familiar management territory; acceptable methods have been evolved for dealing with them.

By contrast, benefits are not always easily identified. Some benefits are 'intangible' in that designers and clients alike may feel that they cannot determine a value for them and on occasion, though clearly felt, they cannot be precisely articulated. Most benefits derive *after* the completion of projects; their sources and extent are not necessarily as anticipated. Benefits perceived are closely linked with expectations and experience. Thus unexpected benefits in unexpected areas of a client organisation or in the marketplace may not be readily associated with work on a design project unless a clear, direct link can be established. Should a combination of factors be involved, there is a tendency to consider that the benefits would have emerged regardless of work on the project. Furthermore, benefits can be open to wide interpretation. What may appear to be a worthwhile benefit to one manager may be considered an unnecessary luxury by another. A range of benefits may fulfil the expectations of some managers, while other managers may judge them as inadequate returns on investment.

Thus, unlike costs, many benefits are not seen as 'real' but imputed and indirect, to be viewed with suspicion. Because they may not be immediate they are uncertain; the longer the time-lapse before a benefit accrues, the greater the uncertainty.

The implications to design project evaluation are several. Normally it is easier to draw up a list of project costs than it is to list the benefits derived. There is a tendency to place greater confidence on costs than on benefits: for while costs appear certain, benefits may never actually materialise. The greatest weight will be attached to benefits which accrue during a project or very shortly after completion. This is only partly due to the higher degree of confidence attached to immediate benefits. There is also a psychological reason: the greater the benefits achieved during or immediately after a project, the greater the perceived success of the project and the greater acknowledgement the project manager receives.

Because most costs arise during the course of projects, managers can and do react speedily to variances in costs. By contrast, as most benefits are enjoyed only after completion of projects, changes take longer to determine and cannot be acted upon with the same speed. The further the changes are into the future, the less credence is attached to them and the less urgent they are perceived to be. Thus many projects are managed with only nominal consideration of benefits, and short-term benefits at that. Firm action will be taken to check cost variances, sometimes to the extent of dropping projects, without reference to the accompanying changes in potential benefits. Often, managers will have clear guidelines on acceptable cost levels for particular types of design project but no similar guidelines on cost/benefit ratios. Consequently there is always a danger that attempts at evaluation become short-term and cost-heavy.

Few organisations have the information or procedural systems to evaluate design projects in the longer term. Any assessment of project outcome usually takes place at the end of each project. The costs, having been incurred, are real and specific; the benefits are interpreted as perceived. All assessments are heavily influenced by expectations. Problems or costs which arise *after* this stage — such as maintenance work on a substandard solution or remedial work resulting from faulty implementation — may not be dealt with by the original project team or within the original project budget. Documentation on these costs may not even reach the project file, if such a file exists. As communications can be inefficient, memories short, and the time-lapse between similar projects long, these costs may never feature in any formal assessment of the project.

There is a corresponding shortcoming where benefits are concerned. If benefits arise after a period of time during which evaluation is considered reasonable, the project file will have been closed and the benefit probably goes unrecorded.

Perhaps the greatest hurdle in drawing up a fair list of project benefits is that, often money must be spent to determine the true extent of benefits. For example, sales figures alone do not provide adequate information on customer and retailer reactions to a redesigned product or an item of promotional literature. Market research is needed to gain detailed feedback. Many managers will not allocate funds to such exercises, preferring to rely on their own judgements and the reactions of colleagues. Some reason that if the outcome of a project is more favourable than their analyses suggest, then that is fine because the benefits are being enjoyed anyway; so why spend money finding that out? Others argue that the increased accuracy in the range and assessment of benefits will not justify the additional expenditure. A few managers will admit frankly that they have no interest in determining the wider significance of design projects: as long as the objectives set are achieved, it really does not matter if the project has wider impact. Though they agree that further benefits are unexpected bonuses, these should not be included in any evaluation.

Whereas it is readily accepted that undertaking project work will cost money, it is less obvious that money has to be invested in order to reduce expenditure in the longer term. Apparently few managers ask the question 'How much does doing nothing cost us?' when seeking project opportunities. Thus the fact that a product is not updated regularly with minor modifications may lead to a substantial loss in sales. The costs of redesigning and relaunching the product will probably far outstrip the 'saving' in keeping the product unchanged. Unfortunately the fragility of such 'savings' — or, rather, the extent of the hidden costs — is rarely grasped early enough, either when projects are planned or during their progress. Thus they do not feature in evaluations. This point relates not only to whole projects but also to parts of projects. For example, the 'savings' effected by cutting out tests on consumer reactions to the redesign of a product may fade into insignificance if the product is a failure.

Control of costs and benefits through variance analysis

Drawing up lists of project costs and benefits and comparing them is only the first step in evaluation. We need to determine how efficiently these results were achieved: that is, how effectively the project was managed.

As is known from basic cost accounting, historical costs are of limited value when analysed in isolation because they provide an inadequate basis for the interpretation of efficiency of performance. One system of cost control involves the comparison of actual costs with predetermined, 'standard' costs in order to isolate variances which help pinpoint deviations from the anticipated course of events of performance, hence possible inefficiencies. 'Standard' costings may, perhaps, be impossible to calculate for design projects, except those which recur regularly. Nevertheless, the concept of control and evaluation through variance analysis can be usefully adapted.

For the purpose of this discussion, cost variances are differences between actual costs incurred during a project (i.e. historical costs) and the costs forecast before the start of a project (i.e. predetermined costs). Sensible forecasts of expenditure can be made at the start of most projects from the basis of accumulated experience with a wide range of projects; they may be agreed between client and designer, or they could be calculated by the client before the designer is introduced to the project. In essence, such forecasts make up the project budget.

From what sources do cost variances arise in design projects? Clearly it is not always possible to forecast accurately the work content of design projects. This may be because the project problem cannot be accurately defined before work starts on its solution. Therefore the project is set up on the basis of cost guesstimates which, being part of the project proposal, may be looked upon as 'standards'. Ill-defined problems often lead to abortive work which naturally adds to costs. Abortive work can also stem from changing priorities as a project progresses. Such changes may require revisions to work programmes: perhaps activities planned in detail for the latter stages of a project need to be brought forward and executed within a crash programme. Dismantling existing arrangements and alterations to contracts awarded will normally add to project costs. Crash programmes invariably raise costs because of overtime worked, the need for extra labour, and higher prices associated with faster delivery of materials often in short supply. Crash programmes are also more difficult to manage efficiently. Even if a problem is accurately defined, the work involved in solving it may not be easily estimated.

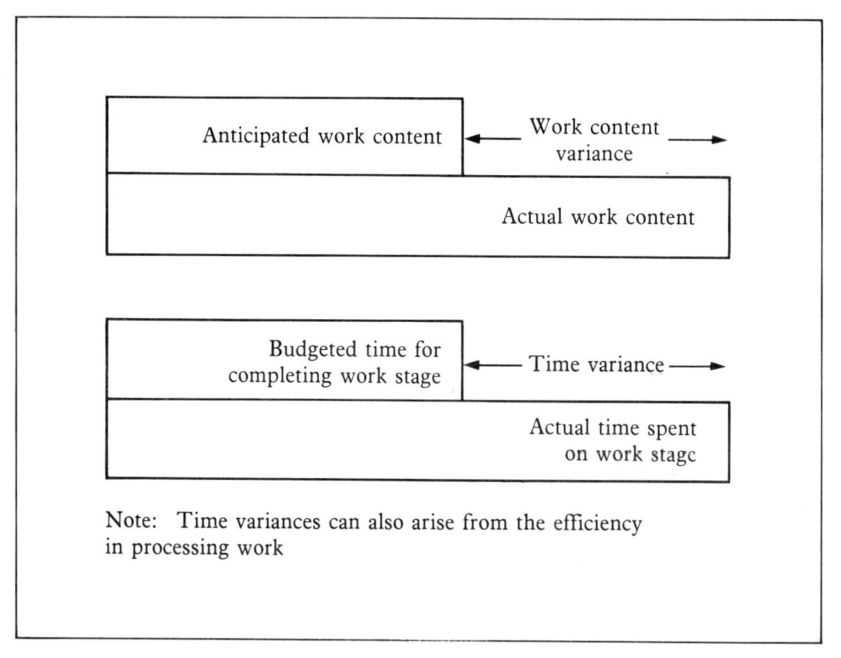

Figure 8.12 *Work content and time variances*

Setting out the work content of a design project is only the first step in the estimation of costs. Elements of work or sections of the project work programme must next be costed out. Further common sources of cost variance are to be found in consultancy fees and personnel charges within the client organisation, material and processing costs, and expenses.

Fees are often expressed as time costs. Thus if a client budgets the work at an hourly rate of £8 and the designer appointed charges £15 per hour, then clearly a cost variance will arise. Many managers underestimate the costs of design work because they are not conversant with going rates.

The level and range of skills used also affect cost estimates and may, in turn, become sources of variances. Certain skills command higher

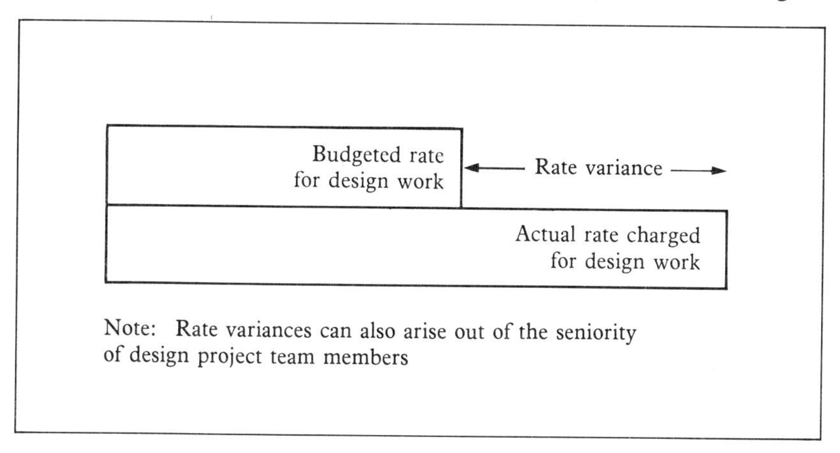

Figure 8.13 *Rate variances*

rates than others; seniority and greater experience will also command higher rates. The more complex and specialised the work, the larger the number of designers, specialists and administrators likely to be used, resulting in higher costs. However, there is usually a trade-off between the level and range of skills used and time spent creating a solution. That is, a more experienced designer may well solve a problem faster and more effectively than a junior designer. Employing a specialist to help out on certain aspects of a problem could also save time. The benefits gained from such time-savings often outweigh the costs of upgrading skills or introducing additional members to the project team.

Care must be taken when calculating 'people' costs, because methods of charging vary among specialists and suppliers. Some quote 'all-in' charges, while others break down their involvement by area worked, perhaps giving different rates for different areas. In the former case there is the danger that the client may think that certain aspects of work are not being charged for, and this may distort estimates for future design projects. Take the example of a printer who does not declare a separate cost for designing or preparing artwork for a particular item: one charge is made for supplying a quantity of these items. As many managers look upon printers as 'producers', they sometimes reach the false conclusion that no charge is being made for design and artwork. Therefore, the use of designers who naturally charge for such work appears to them as an extravagance. In reality, they may be paying *more* for design and

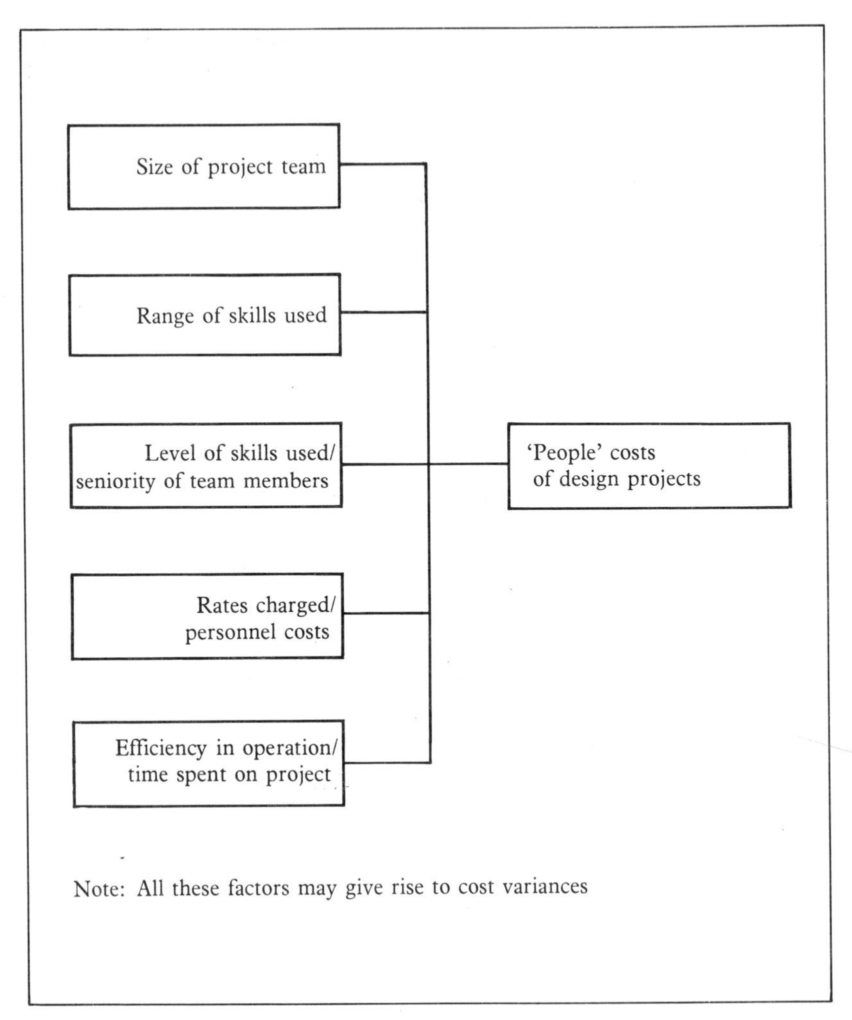

Figure 8.14 *The 'people' costs of design projects*

artwork than they would were a designer involved, but because the cost is 'hidden', the fact is not readily apparent. Neither is it readily apparent that by introducing a designer into such a project, the cost estimate may merely be reshuffled.

Supplies, materials and processing costs may be linked. Designs can be produced which make use of specific processes; these processes might suggest or dictate materials. Similarly, designs may be produced to take advantage of certain materials which can only be manipulated efficiently with particular processes. The careful selection of materials and processes in order to keep costs to a minimum, consistent with the objectives of a project, is fundamental to effective design. Yet

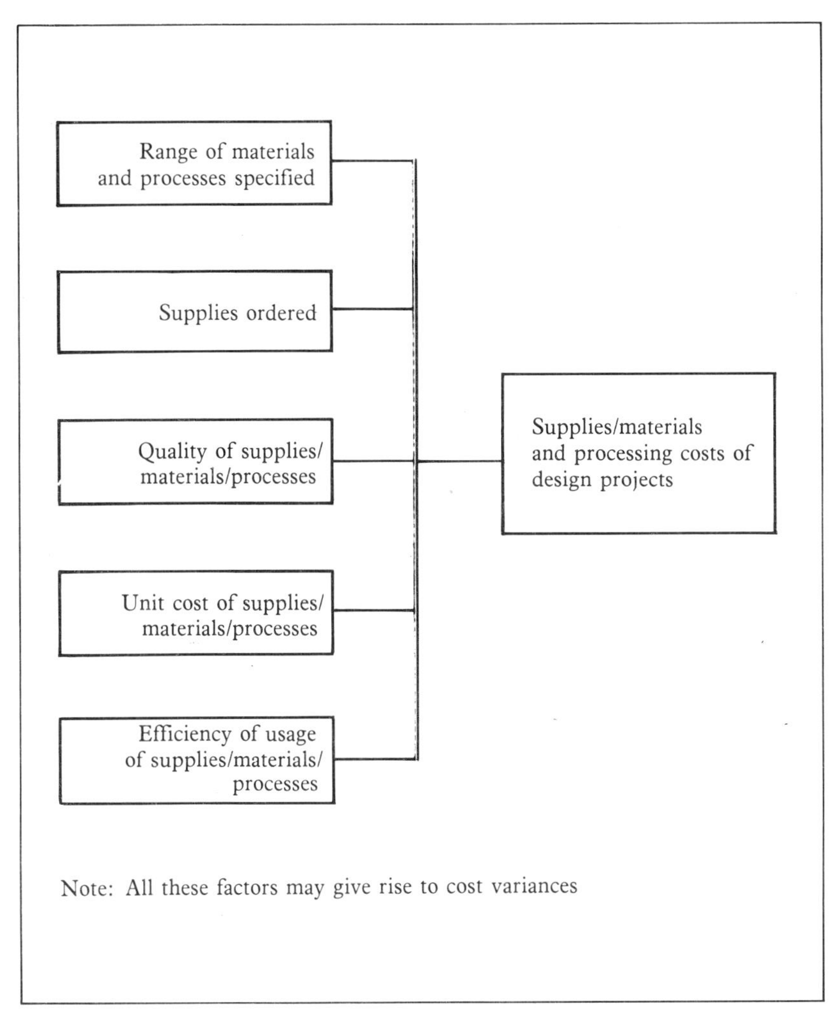

Figure 8.15 *Supplies, materials, processing costs of design projects*

frequently clients will produce estimates based on the use of unsuitable materials and processes. On occasion, materials will be overspecified and expensive processes dictated; at other times, inadequate materials and low-quality processes will be specified. It is not uncommon for the advantages of using quality materials to be nullified by inappropriate processing, or for money to be wasted on the expensive processing of substandard materials. All these circumstances could yield positive or negative cost variances. Again, there is usually a trade-off between the materials and processes used in design projects and the costs incurred. Upgrading to higher-quality materials and more efficient processes can so reduce time and material consumption that costs are actually reduced in the long term.

Expenses are often a major source of cost variance. Managers commonly assume that expenses in design work are negligible, or insignificant compared with other costs, such as design fees. Yet expenses can surpass fees, especially if extensive travel is involved during the course of a project. Again, all-in quotes may provide a distorted view of the proportion of a project budget which is actually spent on expenses. Thus the total cost variance experienced during a design project will be the product of the variance in work content and the variance in costs of the work content.

Through this kind of analysis divergences from planned action and anticipated outcome can be isolated and explained. The questions to be answered are: Where did variances occur; and, why did they occur? Of course, not all variances can be controlled. However, for those variances which are controllable, a further question should be asked: Who was responsible? The answers should provide pointers for remedial action and for improved preparation towards, and management of, future projects.

Cost variances have been discussed in some detail as these are the variances that managers normally look for, though seldom use objectively, in their decision-making. However, it should be emphasised that an examination of project variances would be incomplete unless a corresponding *benefits variance analysis* is undertaken — precisely because many cost variances, whether favourable or unfavourable, are directly related to variances in benefit.

Evaluation and management action

Evaluation is no panacea for the troubles of the design profession or of management. Neither can it be a tool by which design is reduced to a 'numbers game'. Evaluations which do not provide the raw material for action are sterile. In order to encourage industry to invest more extensively in design, managers must be offered the opportunity to compare rigorously an investment in design with the many other investments that compete for limited resources. The process of evaluation will facilitate this kind of comparison if it yields accurate information on which managers and designers can make decisions. This can only be achieved if a long view is taken of design projects (in the sense of looking at repercussions in the further and nearer futures) as well as a wide view (in the sense of allowing for side-effects of many kinds, both inside and outside the client organisation).

With knowledge derived from comparison, the investment decision is extended from one simply committing or withholding resources to one that includes considerations of efficiency of investment and how such investments can be managed.

This discussion has dealt with design project evaluation on a very basic level, with consideration of just *some* of the factors involved.

Clearly there is no one system by which all design projects can be evaluated effectively, just as there is no one system which would be suitable for all business organisations. Ultimately, individual companies will need to devise their own evaluation systems, having considered all the factors involved. Surely what is required is a clearer understanding of the nature and benefits of design projects deriving from a wider base of reliable data. It will only be through such understanding that design is accorded its rightful place in the mainstream of profitable economic activity.

Appendix A
'Design projects are difficult to manage because...'

Managers often express their opinions of design and designers. Design spokesmen frequently take the opportunity to comment on industry and management. Periodically, the design profession will also debate charges — emanating from within and outside the profession — that its members are insufficiently commercial in outlook, that the work of so many designers demonstrates little understanding of economic realities. By contrast, there appears to be no data on how managers see their own roles when handling design projects, and on what they see as the difficulties in managing such projects.

In 1976 a questionnaire survey was conceived to generate data on management or, more broadly, client perceptions of the difficulties encountered when managing design projects. As far as has been ascertained, such a survey has never been carried out before. The survey was conducted in the United Kingdom and Canada over the period February 1977 to April 1978.

Respondents were asked to indicate agreement or disagreement with 28 statements on difficulties perceived in managing design projects. Virtually all these statements were distilled from those typically raised by managers and designers when discussing design. Respondents were also asked to indicate their position in the company, and whether they had been responsible for, or had participated in, design projects. For the respondents' reference, definitions of product, graphic, exhibition, and interior design projects were given on the questionnaire.

The sample numbered 242 managers/clients, 172 drawn from the United Kingdom (71 per cent) and 70 from Canada (29 per cent), covering the full range of management seniority. 73 per cent of respondents had either been responsible for, or had participated in, design projects; most had experience of more than one category of design project.

A sample of 242 may be considered small to draw reliable conclusions. However, analyses on the ten sub-samples have revealed a surprising consistency in response profiles: the significant statements are almost identical in each sub-sample, and the profiles of experienced and inexperienced respondents show similarities. Thus there may well be a common mode in 'management' thinking.

The findings of this survey should be interpreted strictly as indicators of contributory factors (or otherwise) to difficulties in managing design projects. There is reason to believe, nevertheless, that more than a few respondents considered many of the statements put to them as inherent characteristics of design projects.

The full list of 'difficulties' put forward to respondents is given in Table A1 (p.134). Significance in the findings is ascribed to those difficulties which drew *at least* two times the response one way rather than the other.

A second stage of this 'management' survey was launched in September 1978 through which the significant findings emerging from the original survey are to be checked and probed in greater depth. This stage will be undertaken by means of interviews as well as a more detailed questionnaire.

A complementary survey — which seeks to determine *designers'* perceptions of the difficulties of managing design projects — was run on delegates at the International Congress of Graphic Design Associations (ICOGRADA) in Chicago, and subsequently on the designer readership of *Design* magazine in the United Kingdom.

Respondents were asked to agree or disagree with 48 statements on difficulties perceived when handling design projects. In all cases, designers were expected to respond on the basis of their own professional experience — *not* on behalf of all designers. Respondents were also asked to indicate the nature of their professional experience; the number of years in design practice; whether they were with independent design groups, or freelance, or in staff positions with other companies; the categories of project they had been involved in.

Table A2 (p.136) sets out the findings of this survey which derive from 268 respondents of whom 200 practise in the United Kingdom (75 per cent) and the remainder mainly in North America (23 per cent). Just over half the respondents were either members of independent design groups or worked on a freelance basis. The other respondents all held staff positions in a range of organisations. There was a representative spread of professional experience: respondents included, at one extreme, a recent graduate who has just started his first job and, at the other, a senior engineering designer with 40 years' experience. Virtually all respondents had been involved in more than one category of design project.

These findings suggest that there is substantial common ground in the perceptions of designers, whether they work for independent design groups, on a freelance basis, or in a staff position in industry or elsewhere. There is also common ground between designers who work in the United Kingdom and North America. Unfortunately, the sample to date is not large enough for tentative conclusions to be made on the differences in perceptions between designers with varying years of experience.

Taking the 'management' and 'designer' surveys together, it would seem that the common perceptions of designers and their clients indicate that the difficulties encountered during the course of design projects would not be dramatically reduced if designers were to improve their interpersonal or management skills. Most difficulties encountered centre on clients: they suggest restrictions on the potential effectiveness of designers, distractions from the designer's central job, as well as a general lack of application and support on the part of clients. It could be argued that some of these difficulties result from pressures which managers face day-to-day in their work, and therefore designers will just have to learn to cope. But others clearly have their source in management ignorance of the nature of design projects and the work of designers — difficulties made all the worse because the damage is done well before the designer becomes involved.*

Should the findings of these surveys be confirmed through further research, they will provide strong support to those who believe that any design management educational drive should concentrate on managers, not on designers, as the principal target.

* **References**

'Design projects are difficult to manage because. . . ' Report on a survey of management perceptions, July 1978.

'Design projects are difficult to manage because. . . ' Report on a survey of designer perceptions, July 1979.

Table A1 Response profiles of experienced and inexperienced respondents in the UK and Canada, shown as agree/disagree ratios

Design projects are difficult to manage because...

	Experienced		*Inexperienced*	
	UK	*Canada*	*UK*	*Canada*
To manage them effectively requires getting involved in a wider field than indicated by the stated problem	8.0	10.5	13.0	4.3
Senior managers rarely appreciate what design projects involve	4.8	1.1	2.2	1.8
Management does not prepare itself rigorously to carry them out	3.7	2.4	2.6	1.4
They fall, by nature, in unfamiliar management territory	3.6	2.9	2.4	1.4
There is a significant lack of information at the start of each project	2.2	2.3	2.4	2.8
Frequently too much is expected of design projects	1.9	2.1	4.1	2.8
It is difficult to ensure that associated activities necessary to support the solutions of such projects are carried out effectively	1.5	1.3	1.5	1.8
It is hard to select the right designer for the problem at hand	1.4	2.9	5.1	17.0
The way designers work conflicts with the way management works	1.4	0.8	1.4	4.0
Few managers take any real interest in design projects	1.2	1.6	1.6	1.9
Designers are insufficiently experienced in business matters	1.2	1.3	2.2	0.8
Designers have insufficient understanding of business matters	1.2	1.1	1.1	0.9
Design projects evolve, more often than not, from 'cosmetic' needs	1.1	0.8	1.4	1.0
Their results cannot be properly evaluated	1.1	0.8	1.1	0.8

Design projects are difficult to manage because . . .

	Experienced		*Inexperienced*	
	UK	*Canada*	*UK*	*Canada*
Too many unnecessary people get involved	1.1	1.4	1.2	1.9
There is generally inadequate finance to undertake design projects effectively	1.1	2.1	1.0	3.4
It is hard to define design problems	1.0	1.4	1.3	1.9
Senior managers rarely take decisions when necessary	1.0	0.5	0.7	0.3
Design solutions are essentially based on subjective factors which cannot be adequately expressed	1.0	1.3	1.5	0.7
One has to rely so much on the guidance of the designers involved	0.9	1.4	1.4	1.6
Designers do not fit comfortably into business teams	0.9	0.7	1.6	0.7
Designers do not research the client situation in sufficient depth	0.8	0.6	1.5	0.9
It is hard to brief and supervise designers	0.7	0.9	1.0	1.2
There is generally insufficient time to produce effective solutions	0.7	1.1	0.8	1.1
One rarely knows what has been achieved	0.6	0.9	1.1	1.1
They derive frequently from crises	0.5	0.9	0.6	0.5
Design projects require an inordinate amount of resources, such as manpower, materials, and finance	0.3	0.3	1.0	0.5
The business objectives sought can rarely be achieved through the solutions of such projects	0.3	1.2	0.6	1.9

Note: 'Significant' agreement with the difficulties put forward is indicated by ratios of 2.0 and over; 'significant' disagreement by ratios of 0.5 and under.

Table A2 Response profiles of designers practising in the UK and elsewhere, shown as agree/disagree ratios

Design projects are difficult to manage because . . .

	UK	*US+*
To manage design projects effectively requires getting involved in a wider field than indicated by the stated problem	18.0	12.0
Frequently the designer has to educate the client in design management as well as get on with the project	16.7	32.0
Senior managers rarely appreciate what design projects involve	8.1	3.6
It is difficult to ensure that associated activities necessary to support the solutions of such projects are carried out effectively	7.8	5.0
The designer's workload is never predictable	7.4	2.3
Usually the person given responsibility for the project day-to-day is not the final decision-maker in the client company	7.2	13.0
The scope of projects frequently expands as they progress, while deadlines and budgets do not	6.8	12.2
Clients tend to offer pre-digested problems	6.4	5.0
Clients often fail to grasp the wider implications of design solutions	6.2	12.0
There is a significant lack of information at the start of projects	6.1	5.5
Management does not prepare itself rigorously to carry them out	5.9	14.5
The client frequently does not understand his problem	4.8	5.5

Design projects are difficult to manage because...

	UK	*US+*
Often inadequate budgets are set	4.4	5.9
Clients tend to take a short-term view of their problem	4.1	3.3
Frequently designers have to 'design down' to the client's business and marketing competence, rather than to the end-consumer's acceptance	3.1	6.6
They derive frequently from crises	2.7	5.2
Clients often appoint the wrong designer	2.7	1.9
Too many unnecessary people get involved	2.6	3.5
Senior managers rarely champion design work within their companies	2.4	2.0
It is difficult to cultivate an accepting climate within client organisations for 'different' solutions	2.1	1.8
Designers and their clients tend to speak very different languages	2.1	2.4
Design projects evolve, more often than not, from cosmetic needs	2.0	3.4
Clients do not believe designers can assimilate and use business information	2.0	2.2
There is generally insufficient time to produce effective solutions	1.9	1.8
In several instances, the client needs a business consultant. Nevertheless, he appoints a design consultant	1.7	1.0

Design projects are difficult to manage because...

	UK	*US+*
Designers do not always check the relevance of their solutions to their client's business	1.7	2.7
The client and his staff rarely make their full contribution to the project	1.6	1.0
Designers do not research the client situation in sufficient depth	1.4	1.8
Frequently too much is expected of design projects	1.3	1.7
Clients like to get too involved in the design process	1.3	2.8
Their results cannot be evaluated	1.3	0.3
Clients tend not to believe there are effective design solutions to their business problems	1.2	1.2
The designer's objectives tend to be, by nature, different from those of the client	1.2	1.2
Designers cannot deal with client problems effectively on a project-to-project basis as there is no continuity in the client/designer relationship	1.1	0.9
Clients rarely take decisions when necessary	1.0	1.0
It is difficult to contain design problem-solving within deadlines	1.0	1.3
Designers have insufficient understanding of business matters	0.9	1.2
Few managers take any real interest in design projects	0.9	0.5

Design projects are difficult to manage because...

	UK	*US+*
Clients tend to perceive designers as being unsympathetic to business/marketing considerations	0.9	2.8
The way designers work conflicts with the way management works	0.8	0.9
Designers are frequently poor project managers	0.7	0.7
It is hard to define design problems	0.6	0.5
Design solutions are essentially based on subjective factors which cannot be adequately expressed	0.6	0.3
Designers do not make sufficient effort to keep clients in touch with progress	0.6	0.9
Designers do not fit comfortably into business teams	0.6	0.7
Designers rarely grasp the wider ramifications of their client's problem	0.5	0.3
Designers tend to have insufficient skills to deal with clients on an equal basis	0.5	1.0
Designers tend to pay insufficient attention to the client's views	0.4	0.5

Note: 'Significant' agreement with the difficulties put forward is indicated by ratios of 2.0 and over; 'significant' disagreement by ratios of 0.5 and under.

Appendix B
The 'mechanics' of design projects

A recent survey of management perceptions of the difficulties encountered when handling design projects revealed that managers – or, more broadly, clients – consider design projects to be *by nature* unfamiliar management territory (see Appendix A). This is an interesting finding, given that *all* outputs produced by organisations are designed and that managers are responsible for the production of these outputs.

Experience confirms that relatively few managers are conversant with the 'mechanics' of design projects: the typical sequence of events which constitutes progress from the initial awareness of a problem through to the evaluation of an implemented solution. Of course, no two design projects are identical. Yet the similarities between projects – and particularly those within categories of design – are often greater than the differences. Thus projects concerned with the design of interiors for retail outlets will follow broadly similar sequences of events; projects concerned with the design of corporate brochures will follow broadly similar sequences of events, as will projects concerned with the design of products, and so on.

Experience also confirms that though clients may complain it is difficult to keep designers to deadlines and budgets, few lay the foundations for strict project control at the start of projects.

Despite the fact that clients believe they should instruct designers to carry out their requirements, in practice their behaviour is very different. More often than not, clients will provide their designers with bare outlines of problems, expecting them to fill in the detail as necessary. These outlines do not constitute briefs, neither do they constitute programmes of work. At the briefing stage the clients' principal concern tends not to be for the 'mechanics' or the efficient processing of projects, but rather to ascertain whether the designer has 'grasped the essence of the problem'. All too frequently this means little more than ensuring that the designer understands the type and visual style of solution required.

Unfortunately, project management decisions made on the basis of the bare outline of a problem are often inadequate when the full extent of the problem is determined. If a manager does not know the likely sequence of events during the course of a project, he is less likely to plan the work efficiently. In particular, he may make premature decisions fixing deadlines and budgets which seriously affect the chances of producing effective solutions. To be fair, many designers do not inform their clients of the steps involved in working towards solutions. They may agree deadlines but they will not set out a programme of work as a reference. Some designers consider it unnecessary to provide clients with that kind of detail while others have never analysed their work in that way before. Some may genuinely not know how they are going to proceed with that particular project, while yet others consider that agreeing a sequence of events with the client may restrict the flexibility of manoeuvre during the project.

What few designers and their clients realise is that agreement on a plan of action sometimes constitutes the only common ground between them at the start of a project. This may especially be the case when the client has never worked with a professional designer before, or when a designer and client are working together for the first time. The client may have little idea of what his problem is; he may also have little understanding of visual matters. The designer may know nothing of his client's business or the industry in which it operates; he may never have tackled that kind of problem previously. Yet they both need a framework with which to proceed: the designer needs to structure his approach to solving the problem, while the client needs to have some idea of what his resource commitments will be to the project. Agreement on a programme of work provides a *common* framework for both parties.

The programme of work is also valuable because it sets out the respective contributions of client and designer in the quest for a solution: in addition to being a plan of action, the programme indicates those sections to be undertaken by the designer, those by the client, and those to be undertaken jointly. All these details help ensure that there is a sensible division of labour between client and designer, that efforts are coordinated, and that work proceeds efficiently. Mapping out work in this manner provides a useful yardstick against which to manage and evaluate design projects. It should also help boost confidence in the work being done and in the potential of the designer/client relationship.

This appendix lists typical sequences of events which occur in the course of various categories of design project. The sequences are all set out in the three broad phases of design projects: the pre-project, the project, and the post-project phases. However, the degree of detail included in them varies, both between sequences and between phases in a given sequence. Furthermore, as the division of labour between client and designer tends to be negotiated, variations will occur from project to project. Therefore, apart from certain obvious instances, no indications are given of which party should undertake specific events or whether there should be a joint involvement.

The sequences will apply with little adjustment to projects undertaken by in-house designers rather than design specialists external to the client organisation. They should be taken as frameworks for reference, to be amended to suit the particular circumstances of individual projects and organisations.

Typical progress of an office interior design project

Pre-project

Record the cause(s) that drew attention to the need to set up this project.

Analyse organisation's current business, areas of interest, and whether any changes are anticipated in foreseeable future

Analyse the organisation's current office configuration, both as a 'people' system and an 'activities' system:
— what is done within the office
— how do these activities relate
— who carries out these activities
— how these people interact and the patterns of circulation
— what equipment is used for each activity
— what other facilities are needed
— do any of these activities or equipment need special environmental conditions
— how do these activities vary as the business changes
— are any changes in activities and staffing anticipated in the foreseeable future?

Establish what space is needed — currently and in the foreseeable future — for:
— operations
— equipment
— presentations/social functions
— storage
— staff restrooms, catering, etc.

Determine 'workshop' and 'image' areas

Prepare descriptions of the kind of work atmosphere required, together with an indication of the preferred visual style

Determine whether the present premises can be used more efficiently; should certain activities or space requirements be housed elsewhere? Can the present premises be augmented with space nearby? Or are new premises needed: if so, should new premises be bought, leased, rented, etc?

Establish timescale and guideline budget for project — probably worked out on '£ per square foot' basis

Incorporate all this information into a preliminary brief

Discuss project with various designers, seek project proposals

Appoint designer or design team on the basis of the preliminary brief and/or the designer's proposal, or on a negotiated brief with work programme and preliminary budget

Project/pre-site

Analyse client requirements in greater depth, perhaps discuss problems of particular departments/sections with staff involved. Canvass staff opinions of present workings of office/possible improvements; check requirements and go behind surface of the brief

Analyse findings and discuss with client those findings which suggest adjustments to the brief

If new premises are being sought, draw up specification for premises required and circulate to relevant agencies

If existing premises are to be retained, or when new premises are secured, carry out a detailed survey of physical configuration/services

Prepare layout of premises and agree with client. If necessary, discuss layout with principal staff members

Develop design concept on the basis of the agreed layout. Present design concept by means of:

- colour perspectives depicting areas of special interest
- model where a three-dimensional representation of spaces and volumes will help the client come to a decision
- sample board(s) of finishes and furnishings giving an indication of proposed colour scheme
- sample board(s) of fixtures and fittings
- samples or photographs of furniture proposed

Discuss scheme with two or three possible contractors

Draw up preliminary costings for the project

Draw up list of other specialists and consultants required during the project such as quantity surveyors, structural engineers, heating and ventilation engineers, etc.

Enter into preliminary discussions with landlords, planning authorities and fire officers

Work out probable programme of implementation and phasing of work

Obtain client approval of scheme

Obtain client approval of, and appoint other specialists and consultants

Carry through and finalise design detailing for accepted scheme

Prepare working drawings and specifications

Go out to tender

Receive and check over tenders; check delivery dates

Submit tenders to client and recommend a main contractor, or several contractors

Appoint the main contractor, or several contractors

Obtain all necessary planning, landlord, and fire authority permissions

Programme work on site in detail

Communicate what is going on to client staff; indicate how they will be affected and when

Arrange alternative accommodation where necessary

Negotiate and/or firm up budget

Place orders where necessary

Make applications for provision of electricity, telephones, switchboards, gas, water, etc.

Check all insurances have been taken out

Project/on-site

Brief all parties involved in site work and ensure effective liaison between all these parties

Confirm programme of work on site

Complete further detail drawings as necessary

Check all deliveries to site, and approve suppliers' invoices to be paid

Check progress and workmanship on site at periodic site meetings

Monitor all abortive and additional work; record reasons why this work arose

Plan for occupation of premises

Check and have officially passed services and works before hand-over. Draw up schedule of deficiencies and defects. Instruct relevant parties to correct deficiencies and defects at convenience of client

Move staff into premises

Arrange photography of final scheme

Hold hand-over meeting

Provide client with appropriate sets of drawings, specifications, lists of manufacturers and suppliers for reference and maintenance purposes

Post-project

Monitor use being made of premises and compare with pattern of usage anticipated. Where appropriate, suggest remedial action

Monitor office staff and visitors' reactions to new design (layout, visual environment, etc.)

Make final check of premises at end of maintenance/defects liability period. Put in hand any necessary repairs and check when completed

Check performance of finishes, furniture and fittings after period of use

Make list of areas that could be revised at a later date

Review design of premises, the administration of the project, and the respective contributions of the staff of the client organisation and designers

Make list of project management points to be taken into account in future projects

Typical progress of a graphic design project involving the production of a full-colour brochure

Pre-project

Analyse the business problem
- why is a brochure required; what is it expected to do; to what use will it be put; how long will it be circulated; how many brochures are likely to be produced?
- who make up the target audiences?
- what messages are to be communicated; what is the priority of communication; what is the sequence of communication of these messages?

What are the objectives set for the brochure; what reaction(s) and action is it expected to elicit from the reader; are different reactions/actions sought from the different types of readers; are there any other criteria by which the brochure will be judged?

What is the scope of the copy to be; what style of copy should be adopted; what should be the structure of the copy? Prepare a preliminary copy brief, and gather together relevant reference material and/sources

What visual material is to be used to complement and enhance the copy; where will this material fit in (photographs, diagrams, illustrations, etc. — whether monochrome or full colour); does any of this material exist or will it have to be commissioned?

Formulate preliminary project brief incorporating an indication of time scale and a preliminary and/or maximum budget

Approach designers; discuss problem and possible project structure

Draw up short-list of appropriate designers, and request proposals of how they would handle project

Appoint designer

Negotiate and agree final project and (where possible) design brief, work programme and budget

Project

Commission copy writing based on agreed copy brief, and provide relevant reference material and/or sources

Check copy for content and style; check reaction to copy content, structure, style, length and how copy is intended to balance with visual material

Given approved copy, reassess the objectives of, and approach to, the visual design of brochure; if necessary, amend project work programme and budget

Where necessary, order translations of copy into foreign languages; have translations checked for accuracy and style

Develop visual concept for brochure and present to client by means of:
- a cover design
- grid for inside pages
- selected typeface(s) and typesize(s)
- a sufficient number of double-page spreads
- colour scheme to be adopted
- recommended stock for inside pages and cover
- if the brochure is to be circulated principally by mail, an appropriate envelope should also be presented, together with details of relevant inland and overseas postal rates

Obtain 'ball park' quote for origination, sub-contracted work, printing and production, based on visual concept of brochure as presented to the client, and preliminary concept of contents — that is, length of copy, number of photographs, diagrams, full-colour work, pages, number of brochures required, etc.

Lay out complete brochure: plan pagination, fit copy complemented with planned visual material

Prepare finished visual/dummy of brochure

Draw up schedule of photography; commission work and art direct if necessary

Draw up requirements for illustrations, diagrams, etc.: commission illustrations and art direct work if necessary. Similarly supervise preparation of diagrams

Mark up copy and order typesetting

Check typesetting and check fit of copy to pagination plan

Draw up base artwork

Obtain approvals for all illustrations, diagrams and photographs to be used in the brochure, together with where they will be placed, and how they will appear in the brochure

Mask and scale up all visual material

Obtain final quotes from printers

Recommend one printer; get approval of his quote, and place print order for the job

Book machine time; perhaps buy in stock in advance

Complete artwork

Submit artwork for checking and approval by client

Go over artwork with printer, carefully checking over all points of the order

Check page proofs from printers for colour balance, inking, blemishes on plates, alignment of pages, correct stripping-in of visual material, etc.

Submit colour proofs to client for approval

Check quality of finished brochures; delivery

Post-project

Test reaction to total brochure presentation — particularly to copy and visual approach

Review progress of project; efficiency of management; accuracy of budgeting; choice of designer; choice of copywriter; choice of printer; choice of approach to brochure design, etc.

Note: The sequence of events during the project phase relates to the use of the lithographic print process which is currently the most widely-used process for printing brochures. The events would be different when the letterpress or gravure processes are used.

Typical progress of signage design project

Pre-project

What factor(s) drew attention to this problem?

What is the problem:
— is it that space or facilities are being used inefficiently?
— is it that new uses are being sought for the premises?
— have changing priorities prompted a review of the space usage and the signage system?

Collate all available data

Check what others are doing in the field

Draw up activities chart (say, follow a representative sample of users from point of entry into premises right through to point of departure).
— Record range of activities which take place in the premises: those done by individuals/those done by groups/activities expected to be done with information assistance and those expected to be done without assistance
— Record the pattern and frequencies with which activities take place, during the day or over a specified period; how do patterns of activities differ from those sought after?
— Record the locations at which these activities take place within the premises, and locations at which these activities should take place
— Record the points at which information assistance is available, and points at which such assistance should be made available

Draw up messages chart
— What messages are being conveyed by available signs?
— How are these messages conveyed: pictorially or by means of text, or mixtures; how clear are these messages (are single messages conveyed per sign or multiple messages); could more information be made available through signage without confusing users
— What kind of signs are already in use (physical construction: hanging, projecting, freestanding/one- or two-sided/colours/sizes/size and type of lettering, etc.); where and/or on what surfaces are these signs located (height/reading distances/obstructions to sightlines, etc.)

Draw up locational chart

Isolate the critical elements and clarify priorities in the design of the signage system; establish functional objectives; isolate constraints and 'givens'; agree on guidelines for evaluation of signage design and project

Determine the principal 'business' problem(s) underlying the project

Incorporate all this information in a preliminary project brief

Approach designers and discuss the problem with them. Arrange site visits and seek project proposals from them. In certain straightforward, highly-prescribed circumstances, a sign manufacturer might be approached directly

Select designer

Project

Undertake further in-depth analysis of problem. Check all user reactions. Confirm and add to details in preliminary brief

Confirm brief

Develop concept of design scheme and present with outline implementation budget

Gain client approval for design concept

Produce prototype which could be tested on site at strategic locations

Review reactions and make adjustments if necessary

Design signage scheme in detail and present scheme with preliminary programme of implementation and an estimate of budget

Obtain all necessary permissions from landlords, planning authorities, and fire officer

Draw up complete schedule of signs needed, giving details of numbers to be produced, location, exact siting, message(s) conveyed, artwork references, size, whether one or two-sided, whether hanging/free-standing/projecting, colour(s), material, method of production, fixing details

Obtain client approval of sign schedule

Complete all necessary drawings and artwork. Prepare all production information and technical specifications

Submit and discuss drawings and specifications with sign manufacturers and contractors to obtain quotes

Discuss production schedule and possible programme of implementation on site

Draw up plan of implementation

Submit names of manufacturer, contractor and programme of implementation recommended together with appropriate quotes. Obtain client approval of quotes and appointment of names parties

Where necessary, submit artwork to client for checking and approval

Place orders on behalf of client

Confirm detail plan of implementation on site

Supervise production of signs and check quality of finish

Prepare site and programme work in detail: will sections of the premises be closed in a prearranged sequence; will work proceed after working hours, or will work proceed during working hours?

Prepare users of premises for programme of implementation; communicate through posters and notices (or even an audio-visual presentation); provide users with additional help during change-over 'introductory' period

Project/on site

Brief all parties involved in site work and ensure effective liaison between all these parties

Confirm programme of work on site

Complete any further drawings which may be needed by contractors

Check all deliveries to site, and approve invoices to be paid

Check progress and workmanship on site at periodic site meetings and inspections

Monitor all abortive and additional work; record reasons why this work arose

Provide client with appropriate sets of drawings, specifications, lists of suppliers and contractors for reference and maintenance purposes

Arrange photography of final scheme

Post-project

Review performance of new signs over an agreed trial period

Review comments made on new signage scheme

Evaluate the design of the signage scheme in terms of efficiency of communication, impact of the image projected, and physical endurance

Draw up a list of points which may be considered when signage scheme is developed at a later date

Draw up a list of deficiencies of signs installed, and instruct relevant parties to correct these at convenience of client

Review the project from the project management point of view, noting particularly the respective contributions of designer and client

Make list of project management points to be taken into account in future signage projects

Typical progress of an exhibition design project

Pre-project

Set out general exhibition strategy of the client organisation specifying:
— exhibition programme for the next two years
— choices of trade and public shows, and
— broad objectives and ballpark budgets for each

Draw up preliminary schedule for particular exhibition

Set specific objectives for particular exhibition:
e.g. to develop market or to explore new market opportunities. Break down preliminary budget set to distil figures for stand design and construction, displays, staffing, and promotion

Choose stand in appropriate section of exhibition and book space

Clarify the messages to be conveyed by means of the stand

Approach designers and seek proposals

Select and appoint designer or design team

Formulate and negotiate final brief and programme of work, fees and preliminary budget (or maximum total budget)

Project/pre-site

Prepare and agree layout of stand

Develop visual concept of the stand design on the basis of the agreed layout and present by means of:
— model
— colour perspectives
— details/samples of exhibition system to be adopted (if any)
— sample boards of finishes, giving indication of colour scheme
— samples or photographs of furniture proposed
— sketches of how special display features might be handled/treated

Discuss designs with contractors and obtain preliminary quotes

Prepare working drawings and draw up detailed specifications

Submit proposed design for exhibition organisers and fire officer for approval

Obtain final quotes from contractors

Make choice of stand contractor. The contractors for most services may well be nominated by the exhibition organisers

Plan pre-site construction and site work

Prepare exhibits and displays

Prepare graphic/photographic displays

Arrange transportation and cranage

Order exhibition system (if any is to be used)

Agree personnel complement and work out staffing roster

Order furniture, carpets, services, etc.

Arrange necessary insurances, accommodation, photography, etc.

Plan advertising and publicity

Design and order operational print, publicity material, promotions

Organise catalogue entries, press releses, back-up advertising. Arrange inaugural reception if one is to be held

Draw up guest list; send out invitations and passes

Project/on-site/erection

Brief all parties involved in site work and ensure effective liaison between parties

Supervise progress of work:
- check all deliveries on site and approve invoices to be paid
- position plinth, carpet on floor slab, install services, erect stand, decorate, install displays and furniture
- check that all work is carried out according to plans and specifications
- check quality of finish
- obtain all necessary official approvals

Clean stand and photograph before exhibition opens

Project/on-site/pull-out

Supervise work during pull-out

Remove loose/freestanding elements; pack and label adequately

Dismantle stand, and pack system (if a system was used)

Ensure that stand site has been properly cleared

Post-project

Given objectives set for the exhibition, analyse the total number of enquiries and level of sales (if relevant). Analyse the number of useful contacts made during the show; estimate the number and type of visitors on the stand

Analyse the stand design from the operational, visual and technical points of view. Review the comments received from staff and visitors.

Analyse operational considerations, particularly convenience of layout, efficiency of space usage, resistance to wear and tear, etc. Analyse visual considerations such as appropriateness as a corporate presentation, visual impact in itself and within the context of the exhibition

Review publicity gained: coverage in the trade/general/design press, etc.

Assess re-usability of parts of the stand at other shows, within offices, or in works

Review efficiency of project management, noting particularly the respective contributions of client company staff and designer

Draw up actionable recommendations on basis of post-project analysis on team structure, exhibition programme, stand design, and project management.

Typical progress of product design project

Pre-project

Record what cause(s) brought the problem to the attention of management

Define problem specifying marketing, production and financial dimensions

Gather all analyses and research findings which led to diagnosis of the problem. In particular summarise the product performance during, say, previous five years or since launch (whichever is the shorter), and review product strategy within the context of client company's overall marketing strategy and long-term plans

Formulate preliminary objectives for project

Collate all information and material necessary to draw up preliminary brief for project, specifically:
- description of product
- place in company range of products
- product features and required functions
- minimum performance requirements under normal/abusive use
- production facilities and skills available
- range of materials and processes which might be used
- components specified
- other 'givens'/constraints on design (perhaps suppliers to be used, etc.)
- quality of finish sought
- visual attributes and style/company product image
- probable volume to be produced
- cost of manufacture to be aimed at
- probable selling price
- distribution channels used
- what retail problems are likely to be encountered, if any

Hold exploratory discussions with appropriate designers or design groups; arrange visits to production plant and retail outlets

Describe the client company procedure for dealing with product design projects

Determine preliminary scope of project: does it encompass minor modifications or a complete redesign? Does it encompass packaging and point of sales material, etc?

Draw up project proposal based on outline marketing and design brief, and submit for approval of preliminary sum of money to be allocated to further consumer/market research and design work

Obtain approval of first stage of project

Select designer or design group

Review and finalise product specification and project brief

Form development and design team to include appropriate specialists and managers. Clarify and agree questions of authority and control within the project team and between the team and others in the client organisation

Project/conception

'Brainstorming' period, perhaps with further inputs of market research and technical/production information relating to client company or of a general nature

Sketch out a number of possible product concepts to be discussed between client and designers

Develop one alternative from concept through to the formulation of a possible solution

Present preliminary scheme to client by means of finishing drawings, some working drawings and block models (where necessary)

Prepare preliminary financial statement, setting out outline costs of manufacture and anticipated benefits from project, etc.

Project/development

Detailed analysis of design proposals undertaken by all members of project team (technical development, production, marketing, marketing research and sales departments, etc.)

Discussion of strong and weak points of proposals; agree amendments, compromises, etc.

Construct block model and/or full working model

Check market and distribution network reactions to design proposal

Review reactions; amend brief and authorise further development work, or freeze design

Draw up preliminary marketing and promotional strategy

Draw up brief for packaging and items of promotional back-up. Commission necessary design work

Carry out more detailed costing of proposal

Submit final marketing/production document to company board, together with proposed design, for sanctioning of expenditure on prototyping, tooling up, etc.

Order and test prototype

Use prototype or model for preliminary work on promotion programme

Complete full set of working drawings

Prepare detailed production schedule including manufacture (materials/components preparation; component manufacture/testing/assembly/retesting/packaging, etc.), buying in, assembly, testing, and so on

Prepare full stock/buying-in schedule

Prepare detailed marketing and promotion programme

Finalise comprehensive financial statements (capital/cash flow) for various levels of manufacturing output

Agree concepts of packaging and promotional items

Order tooling up, components, stocks of materials, machinery, etc.

Carry out trial run of product manufacture and extend to pre-production run in order to de-bug system

Test market on limited scale

Review and finalise marketing and promotion strategy and review production schedule

Finalise pack design and place order for packaging

Post-project

Proceed with production run

Build up stocks of product in distribution chain

Launch product formally with necessary promotional back-up

Obtain market feedback

Carry out short-term review of product performance

Carry out a longer-term review of product performance. Draw up a list of product features which may need revision

Review project from the administration point of view, analysing particularly the respective contributions of the designer and staff of the client organisation. Make list of project management points to be taken into account in future projects

Set up formal project for development/updating of product

Develop/update product

Typical progress of project involving the design of a range of stationery

Pre-project

Record the cause(s) that drew attention to the need for setting up the project

Draw up a list of items in range of stationery and gather samples of each item

Obtain items of principal competitors' stationery, and other 'approved' designs

Prepare brief description of the client organisation, the industry and markets it operates in

Formulate the objectives for the project

Specify the copy to be set out onto the letterheading and indicate priority of information to be conveyed. Ensure that all statutory information is included in this list. Repeat for each item in range

Specify how each item is used

Specify what elements on the existing stationery must be changed and set out necessary constraints to new designs (e.g. company logotype and/or symbol, standard typefaces and corporate colours if any, margins and spacings for machine processing, etc.). Gather artwork and samples of each of these items for designer's reference. Indicate reasons for changes

Gather samples of items which indicate preferred way of laying out letters, forms, references, etc. Details of typewriters used should be noted. If any items are filled in by hand or machine other than a typewriter, details should be collated

Determine pattern of usage of each item in stationery range as well as quantities and frequencies of reorders

Work out deadline for introduction of new designs and allocate ballpark/maximum budget for project (covering research, design and artwork costs)

Incorporate all above information and material into a preliminary brief

Discuss project with various designers or design groups; invite those short-listed to visit company; seek project proposals from them; make choice of designer or design team

Appoint designer/design team on basis of the preliminary brief, the proposal submitted, together with any negotiated amendments. Draw up and agree a work programme

Project

Undertake further analysis of client's particular circumstances, industry, markets, and competition wherever necessary

When appropriate, stop reorders of stationery to run down stocks in anticipation of new designs

Evolve design solution from concept through to formulation. If there is to be a change in the corporate symbol or logotype, this would be worked on first and submitted for approval

Present design concept to client in the form of rough visuals, giving indication of:
— logo and/or symbol design
— colour proposed (sample swatches)
— typefaces and typesizes adopted (printed samples)
— sizes of paper
— layouts
— paper stock recommended (plus samples)

An indication of letter layout should also be presented, perhaps as an overlay

Indications of the costs of printing a specified quantity of each design proposed should also be submitted at this presentation

Obtain client's approval of designs (and letter layouts, etc., if different from systems used)

Obtain client's approval of print quotes

Prepare artwork and/or print specification according to the print process chosen and printer selected to handle the account

Submit artwork to client for checking and approval

Place print order: instruct printers, supervise print work or liaise with printers during printing

Where necessary, submit proofs to client for approval

Check quality of finished work before delivery to client

Post-project

Check whether:
— new designs are being used as instructed (e.g. typing layouts, folding, etc.)
— response to new stationery designs, both within and outside the organisation is favourable
— new design is proving efficient in use
— print quantity was adequate
— any changes are indicated before reordering or at some later date

Review project from the administration point of view, analysing particularly the respective contributions of the designer and staff of the client company.

Draw up list of project management points to be taken into account in future projects

Index